MAPPINGS: Society / Theory / Space

A Guilford Series edited by

MICHAEL DEAR: *University of Southern California*
DEREK GREGORY: *University of British Columbia*
NIGEL THRIFT: *University of Bristol*

APPROACHING HUMAN GEOGRAPHY
An Introduction to Contemporary Theoretical Debates
Paul Cloke, Chris Philo, and David Sadler

APPROACHING HUMAN GEOGRAPHY

Paul Cloke is currently Professor of Human Geography at the Department of Geography, St David's University College, Lampeter, Wales. He is the author of several books and many papers dealing with the rural geography of Western societies, and most recently he has authored *The Rural State? Limits to Planning in Rural Society* (with Jo Little) and *The Political Economy of Rural Areas* (provisional title of book in preparation). He is also leading two ESRC research projects, one examining 'Rural lifestyles' and the other examining the 'Economic impact of new middle classes in rural areas'.

Chris Philo is a lecturer in human geography at the Department of Geography, St David's University College, Lampeter, Wales. His present research is on the historical geography of institutions treating mentally disturbed people, and he has written papers on various aspects of 'asylum' geography and also on the connections (both contemporary and past) between philosophy, social theory and geography. In addition, he is preparing an edited collection on *Selling Places: The City as Cultural Capital* (with Gerry Kearns).

David Sadler is a lecturer in human geography at the Department of Geography, University of Durham (after having lectured in Lampeter, 1987–9). He has written widely on the economic, political and social implications of contemporary industrial restructuring, and recently has authored *The International Steel Industry: Restructuring, State Policies and Localities* (with Ray Hudson) and *A Tale of Two Industries: The Contraction of Coal and Steel in the North East of England* (with Huw Beynon and Ray Hudson). He is presently also preparing a book *Global Region: Production, State Policies and Uneven Development*.

Gregory, Martin, Smith
1991
Rethinking Human Geog

Approaching Human Geography

An Introduction to Contemporary Theoretical Debates

PAUL CLOKE, CHRIS PHILO and
DAVID SADLER

THE GUILFORD PRESS
New York   London

© 1991 Paul Cloke, Chris Philo and David Sadler

Published by The Guilford Press
A Division of Guilford Publications, Inc.
72 Spring Street, New York, NY 10012

Last digit is print number: 9 8 7 6 5 4 3 2

Library of Congress Cataloging-in-Publication Data

Cloke, Paul J.
    Approaching human geography : an introduction to contemporary
theoretical debates / Paul Cloke, Chris Philo, and David Sadler.
        p.   cm. — (Mappings : society/theory/space)
    Includes bibliographical references and index.
    ISBN 0-89862-490-8 (pbk.)
    1. Human geography—Philosophy.   I. Philo, Chris.   II. Sadler,
David.   III. Title.   IV. Series: Mappings.
GF21.C65   1991
304.2—dc20   91-2656
                                                       CIP

CONTENTS

PREFACE

In his novel *The Songlines* (1987) Bruce Chatwin reveals just how important a conception of what he calls 'totemic geography' is to the way in which Australian Aborigines understand both themselves and the world around them. He explains how the Aborigines suppose the landscapes of Australia to be criss-crossed by a 'labyrinth of invisible pathways' known as 'Songlines', each of which follows the path taken by a particular 'Ancestor' – by one of the beings who wandered over the continent during the earliest days of creation, the 'Dreamtime', singing out the names of everything they encountered – and each of which thereby lends significance to the rocks, hills, valleys, rivers and also human settlements through which it passes. And what this means is that an Aboriginal individual hence draws upon his or her knowledge of this imaginary geography, a knowledge rich in detailed information about both specific locations and the relative positions of different locations, in order to make sense of what 'takes place' in their own lives and in the lives of the 'tribal groupings' to which they belong.

It can therefore be argued that the Aborigines possess an abundance of what David Harvey (1973, p. 24) once termed the 'geographical imagination':

> This imagination enables the individual to recognise the role of space and place in his [*sic*] own biography, to relate to the spaces he sees around him, and to recognise how transactions between individuals and between organisations are affected by the space that separates them. It allows him to recognise the relationship which exists between him and his neighbourhood, his territory, or, to use the language of the street gangs, his 'turf' .... It allows him to fashion and use space creatively and to appreciate the meaning of the spatial forms created by others.

Writing in 1973, Harvey was suggesting that, whilst many people and certain professions (including academic geographers, city planners and architects) may possess elements of this geographical imagination, a problem still arises for contemporary Western society because this imagination is *not* all that deeply held or widespread. Indeed, he claimed that many of us possess a 'sociological imagination' – an ability to grasp something of how our individual lives are bound up with the 'larger' workings of society and its long-term historical development – but that few of us appreciate the extent to which our lives in contemporary society are shaped by the geography of its component parts. The Australian Aborigines clearly have such an appreciation, although it is true that their geographical imagination is organised far more around mythical materials and spiritual values than would be the case with our own (which is not to say that myth, spirit and symbolism are absent from the human geography of today's Western world). Nonetheless, the idea that we need to foster a more geographical way of understanding – of 'reading', describing and explaining – the economic, social, political and cultural phenomena of contemporary society is a powerful one, and it is one that inspires the writing of this book.

When Harvey was discussing the geographical imagination in 1973, he was doing so at a time when human geography as an academic discipline had just entered into a period of considerable turmoil. To be more precise, following about fifteen or so years in which a measure of consensus had prevailed regarding the approach suitable to human geography – an approach variously referred to as the 'new geography', 'locational analysis' or 'spatial science' – numerous geographers had begun to question the theories, the methods, the relevance and even the politics associated with this approach. Harvey was one of these critics, and the overall tenor of his 1973 volume, *Social Justice and the City*, was to criticise the impoverished geographical imagination of geography as spatial science, and at the same time to argue for a dramatic sharpening of this imagination through an engagement with Marxist concepts capable of demonstrating the interlinkage of social inequalities with spatial forms. But Harvey was only one voice among many, albeit a particularly inspirational one, and other voices called for the geographical imagination to be (re)invigorated in other ways: principally more humanistic ways designed to capture the active roles played by both researchers (the academic geographers) and researched (the people under study) in experiencing, creating and endowing meaning to space and place. There was considerable disagreement among those geographers seeking to map out a way forward for the discipline, then, but a thread common to most of them was the rejection of *positivism*: a philosophy that makes certain 'ontological' assumptions about what exists in the world and certain 'epistemological' assumptions about how we can know the world, assumptions that are actually very restrictive, and a philosophy that clearly underlay much of the practice of spatial science. As a result it is perhaps not inappropriate to talk about *post-positivist approaches in human geography* (N. Smith, 1979) and we have duly set ourselves the task of addressing the various post-positivist strands now being woven together in an attempt to expand the 'geographical imagination' of geographers themselves, of academics in other disciplines and even of people beyond the institutions of higher education.

Our intention here is principally to provide a clear and concise 'introductory survey' that will amount to a 'portfolio' of approaches taken by post-positivist human geographers. We aim to provide as accurate an account as possible of the various approaches we discuss, and to do this we seek to describe particular approaches very much according to their own distinctive theories, concerns and outlooks on what the project of human geography is all about. At the same time we seek to demonstrate the diversity of opinion and position that is often present within particular approaches, whilst also outlining the 'critical edges' where particular approaches are criticised from without by advocates of other approaches. And underlying these ambitions we endeavour to identify both the personnel and the chronology of differing approaches: isolating the major geographers involved, the philosophers and social theorists who have been drawn upon, and the times at which specific arguments have been advanced and then refuted. Given this scope to our book we obviously hope that it will prove useful to undergraduate students pursuing courses in the theory of human geography (indeed, this book has grown out of a second-year 'Theory' course taught at St David's University College, Lampeter), but we also hope that it will be of wider interest to academic human geographers looking for an overview of trends in post-positivist human geography. Moreover, we believe that at various points in our narrative we provide interpretations of these trends, suggesting connections, criticisms and ways forward that might themselves be seen as contributions to the current 'rethinking' of the geographical imagination. We have tried to be faithful wherever possible to 'original sources', so that we record what Harvey himself wrote (for instance) rather than simply drawing upon what another author has said about Harvey, and this means that the book makes more extensive use of quotations from 'original sources' than is normally the case in introductory texts (and also contains a more comprehensive bibliography). Accordingly, we feel that this book attempts something rather different from most other works dealing with the subject of 'approaching human geography', and we like to think that the book 'fills a gap' between advanced state-of-the-art texts (e.g. Johnston, 1985; Kobayashi and Mackenzie, 1989; Macmillan, 1989; Peet and Thrift, 1989) and those teaching texts that run the risk of obscuring both the originality and the complexity of theoretical materials.

The organisation of the book is quite straightforward, but the neatness of the chapter divisions should not disguise the presence of recurring themes – notably to do with the geographical study of particular regions and to do with the intersection of society and space – that cut across chapters and in effect constitute other 'stories' in the recent history of the discipline. In our opening chapter we summarise the movement of human geography away from its early twentieth-century concerns with human–environment relations and regional syntheses, and into the positivist human geography (geography as spatial science) that became dominant – though certainly not in the minds of all human geographers – during the late 1950s and 1960s. This chapter then charts the reaction this approach to human geography engendered after about 1970, and in so doing both signposts the subsequent chapters in the book and discusses the more personal involvement with post-positivist human geography of specific geographers teaching or being taught during the 1970s and 1980s. In Chapters 2 and 3 we examine the two chief routes geographers took when travelling away from spatial science – namely, *Marxist geography* and *humanistic geography* – and in Chapter 4 we discuss the proposals for a *structurationist human geography* that began to appear *circa* 1980 when it was recognised that the 'structure-orientated' approaches of Marxist geography and the 'agency-orientated' approaches of humanistic geography could perhaps be reconciled. In Chapter 5 we consider the roughly contemporaneous proposals for a *realist geography*, which has 'structure-agency' concerns similar to those of the structurationists, but which also seeks to underpin these concerns with a more rigorously philosophical and methodological account of how the human world is 'constituted' and of how researchers can attain knowledge of this constitution. In Chapter 6 we tackle the very recent suggestions that have been advanced regarding a *postmodern geography*, and here we distinguish between the geographical study of postmodernity as 'object' (as the condition of contemporary society) and the emergence of a postmodern 'attitude' that is sensitive to 'difference' – including geographical difference – and is hence acutely critical of the 'totalising' tendencies present in most theorising about the human world. Throughout these chapters we indicate the way in which the different theoretical materials feed into differing conceptions of space and place, and also of how space and place are bound up with both the workings of society 'as a whole' and the experiences of particular individuals and groupings, and we think that taken together our chapters capture something of what post-positivist theories can bring to the geographical imagination.

We want to acknowledge at the outset that each of us has been primarily responsible for the writing of two chapters – Paul Cloke for Chapters 4 and 5, Chris Philo for Chapters 3 and 6 and David Sadler for Chapters 1 and 2 – although we have all discussed and had inputs to each other's contributions. Inevitably, this arrangement has led to some differences between the chapters in terms of writing style and views expressed, but to some extent variations between the chapters also reflect the nature of the materials under discussion. The politically charged languages of Marxism call for a politically aware treatment in Chapter 2, for instance, but similar arguments can be made too about the more poetic languages of humanism and postmodernism or about the more analytical languages of structuration and realism. Furthermore, it would be misleading for us to pretend that we agree with each other at every turn, and it should not be too difficult for the reader to detect where the sympathies and suspicions of each of us lie (and something that most post-positivist approaches do agree upon is the need for authors not to submerge these sympathies and suspicions beneath a veneer of objective 'scientific' detachment).

We are acutely conscious that in an undertaking such as this we can only provide an outline chart of the waters to be navigated, and that in striving to impose pattern on the many currents and tides we may gain clarity through over-simplification and through imputing logic or direction to approaches where none was apparent at the time. Our account will inevitably suffer from sins of omission, furthermore, and we should immediately acknowledge that our focus upon geographical debates conducted in English must close us off from numerous important contributions made by geographers the world over. Nowhere is our choice of what 'ground' to cover more problematic, though, than in the limited attention we pay to *feminist geography* and to the very real difference feminist arguments are undoubtedly making to the theory and practice of human geography more generally. It is surely the case that these arguments – both where they render women visible as 'objects' for geographical inquiry and where they promote a theoretically informed sensitivity to gender relations (see Hayford, 1974; Women and Geography Study Group of the IBG, 1984; Tivers, 1985; Little, Peake and Richardson, 1988; Mackenzie, 1989; McDowell, 1983, 1989a, 1989b) – are having at least as much impact upon the discipline as the various approaches dealt with explicitly below, and in some instances probably more of an impact, although it might be added that feminist geography is itself influenced by and to some degree fragmented into

Marxist, humanist, structurationist, realist and postmodernist approaches. It will, it is hoped, be apparent that an awareness of feminist geography and of broader feminist arguments is not wholly absent from our book, but we have decided against 'ghettoising' a consideration of feminism into a discrete chapter, in part because we believe that the issues involved are more important than this 'ghettoising' might imply (feminism in geography should surely be seen as more than just 'another' approach) and in part because we feel it more appropriate to leave such a consideration to others better qualified. This is not because we suppose that men cannot meaningfully discuss feminism in geography, but it is to recognise that at present such discussion is most ably pursued by women within the discipline who are developing new ways of theorising and practising a human geography sensitised to gender issues.

A final point that follows on here concerns the fact that – despite differences between the three authors in terms of backgrounds and views, some of which emerge in the book – we are still examining approaches to human geography through roughly similar eyes (male, white, 'middling' in terms of both class and age). One of the most significant arguments now becoming current in both the social sciences and the humanities is that many different 'voices' should be heard discussing their own literature, their own history, their own anthropology, their own sociology, their own geography, and so on. And this means that in human geography it should not just be a case of Cloke, Philo and Sadler writing the geographies (say) of labour, of women, of ethnic groups or of children: rather, it should also be a case of labour, women, ethnic groups and children having a large input into, if not always writing for themselves, studies of their own geographies and, in the process, hinting at the many ways in which they suppose space and place to affect their own lives. To say this is also to recognise certain limits to what can be achieved (and to what should be claimed for) the theoretical approaches tackled below, and to conclude this Preface – and in so doing to move from the Aborigines of Australia to the Maoris of New Zealand – the following is what Evelyn Stokes (1987, p. 121) argues in a piece contrasting the 'geography of Maoris' with 'Maori geography':

> Maori geography is another way of viewing the world, another dimension, another perspective on New Zealand geography. But – kia tupato, Pakeha. Be careful Pakeha [non-Maori person, usually of European descent]. Tread warily. This is not your history or geography. Do not expect that because you are an academic or trained researcher in the Pakeha world that all this will come easily to you. Your degrees, your experience and list of publications may be more of a hindrance than a help.

We hope that we have now made clear something of the thinking behind our book, coupled to some appreciation of both its strengths and its weaknesses, and we would like to conclude with a few acknowledgements to people who have helped us along the way.

Various people have read and commented upon chapters, and in this respect we particularly want to thank Mark Goodwin (who has been a constant source of inspiration and distraction throughout) as well as Derek Gregory, David Ley, Martin Phillips, Jennifer Robinson, David Seamon and Nigel Thrift. Connie Gdula, Trevor Harris and Caron Thomas have helped to smooth the passage from rough jottings to finished manuscript, and to them – along with Phil

Bell, John Crowther, Dave Kay, Graham Sumner and Mike Walker – we owe thanks for encouraging us through this project. Special thanks are also offered to Viv and to Jeanette, who have lived through the writing of this book with forebearance and support. And finally, we would like to thank our students at SDUC, and notably the second year of 1989–90, who endured, made suggestions about and perhaps even learned something from our first attempt at teaching the theory of human geography along the lines of this book.

The authors would like to thank all those individuals who have given their permission for the use of quotes, extracts and diagrams.

Paul Cloke,
Chris Philo and
David Sadler
(1991)

1

INTRODUCTION: CHANGING APPROACHES TO HUMAN GEOGRAPHY

Introduction

One of the most obvious characteristics of contemporary human geography is its diversity of approach. Within human geography today there is an unprecedented liveliness to the engagement with issues of method and theory. Rarely, if ever before, has the subject seen such a plurality of research methodologies and encompassed such a broad sweep of topics of investigation. Yet these developments – certainly prolific, occasionally fruitful – are also frequently difficult to comprehend. Our primary task in this volume is therefore a straightforward one: we aim to offer a series of guidelines to the main approaches by tracing out their shifting currents of thought during the 1970s and 1980s.

In one sense at least this is a good time for such reflection, for over two decades have elapsed since the first stirrings of disquiet with some of the results of geography as spatial science (as opposed to earlier reactions against it) shattered the apparent consensus within geography during the 1950s and the 1960s. Remarkable things have happened in geography in the last twenty years. Whilst the advance of spatial science was marked by the publication of two volumes edited by Richard Chorley and Peter Haggett – *Frontiers in Geographical Teaching* (1965) and *Models in Geography* (1967) – several books appeared at the end of the 1980s celebrating the pace and extent of change subsequent, taking the intentions of these earlier volumes as their explicit starting point. They included Bill Macmillan's *Remodelling Geography* (1989), which aimed to celebrate the advances made in the application of scientific method since the 1960s. In their introduction, Chorley and Haggett (1967) argued that books such as *Models in Geography* had opened up the discipline, shattering a previously shared orthodoxy and, in so doing, creating the space within which an (occasionally bewildering) plurality of approaches could develop in the subsequent twenty years. Some of these were reviewed in two very different texts, *Horizons in Human Geography* (edited by Derek Gregory and Rex Walford, 1989) and *New Models in Geography* (edited by Richard Peet and Nigel Thrift, 1989).[1] Peet and Thrift championed the emergence of a new 'political-economy' approach to human geography; Gregory and Walford sought, like Chorley and Haggett in 1965, to introduce a realm of new developments to those educating the next generation of geographers – the schoolteachers. As academic geography enters the 1990s, however, we feel it important to chart these different strands of debate in a systematic and genuinely interpretative fashion; to draw together some of the diffuse lines of thought in a collection that is accessible not just to the practising academic but also to the discerning and committed student; and both to outline what has gone before, and to signpost possible avenues of future progress, in a way that pays due attention to the discipline's earlier history.

Yet whilst the time would appear ripe for such reflection, in another sense it is very difficult to produce a clear-cut or definitive guide to developments in the last two decades. Many intangible issues to do with theory and method are only now being fully understood and discussed, and many of the approaches considered here are still being fashioned. They are characterised more by a state of seemingly constant revision and, not infrequently, confusion and mutual misunderstanding. The task is one we consider worth the attempt, however, if only to clarify some of the more commonly held misconceptions. We believe that much recent debate has rested upon unduly dogmatic assertions of points of difference, and that this has been to the occasional detriment of many of the continuing threads of geographic endeavour. In a sense therefore our interpretation in this volume rests upon two factors. The first is a desire to explain different approaches critically but sympathetically – evaluating their respective strengths and weaknesses not impartially (for such would be practically impossible) but respectfully, mindful of particular contextual circumstances, opportunities and constraints. And the second, which partly follows from this, is to stress as far as possible the twists and turns to geographic thought: not out of any desire to paint a uniform picture of geographic intellectual progress, but rather to explore the ways one set of assumptions and ideas about geography rests upon (re)interpretation of others. Progress in geography, in other words, is not a function of a series of isolated events, but a much more broad-ranging and inter-related affair. The whole is far more than the sum of the parts.

It is in pursuit of these goals that our account starts in the discipline's early history, charting not so much its classical origins as its emergence as a formal university subject in the late nineteenth and early twentieth centuries. Then in the following section we examine the main lineages of the quantitative revolution and the corresponding rise of geography as spatial science as a reaction against the discipline's older regional tradition. We describe the key moments of this drama, so as to set the stage for the main content of this book. An understanding of one act, in other words, is necessary for the remainder of the play – even if its finale is not yet, nor probably can be, written. In the next section, we introduce the main elements of the approaches that developed largely in opposition to, but partly alongside, the quantitative and spatial scientific revolutions, and that are the subject in detail of the remainder of the book. Finally, we consider a series of important questions that run through the story told here rather like a sub-plot, to do with the role of disciplinary boundaries and academic career paths.

The emerging concerns of geography in the late nineteenth and early twentieth centuries: environmental determinism and the regional concept

Just over a century ago, on the occasion of the founding of the first British university undergraduate geography course at Oxford, the pioneering British geographer, Halford Mackinder, addressed the Royal Geographical Society (founded in 1830) on the scope and methods of geography (Mackinder, 1887). He began by posing the question, 'What is geography?', and gave two reasons why such a concern was of pressing importance. The first was the 'educational battle' (p. 141) then being fought to establish geography in the curricula of schools and universities; in other words, the struggle for the very existence of geography as a recognisable body of knowledge. The second was closely related: for, after fifty years of the explorations and expeditions that had become a mainstay of geographical existence, an end was in sight to the 'roll of great discoveries' (*ibid.*). Thus geography faced a crisis (which of course was also an opportunity) in terms not just of intellectual rationale but also of practical endeavour. Mackinder (*ibid.* p. 143) was confident, though, of a future in which geography was defined something like this:

> It is especially characteristic of geography that it traces the influence of locality, that is, of environment varying locally. . . . I propose therefore to define geography as the science whose main function is to trace the interaction of man [*sic*][2] in society and so much of his environment that varies locally.

On this basis, he argued, 'a geography may be worked out which will satisfy at once the practical requirements of the statesman and merchant, the theoretical requirements of the historian and the scientist, and the intellectual requirements of the teacher' (p. 159).

Mackinder's concern was significant in more ways than one. In isolating the need for geography to identify itself, its methods, and its relation to the world, he presaged many of the debates of subsequent years. Yet he also spoke of issues that were of pressing, indeed vital, significance to the status of geography at that time. For after the death of both Alexander von Humboldt and Carl Ritter in 1859, an era in the history of geography had come to an end (on this earlier, classical, phase, see Glacken, 1967; James, 1972; Bowen, 1981). With the increasing breadth and specialisation of knowledge and education, the subject was forcibly entered into a battle for academic survival, as it sought to establish both a place in the world and a role for its place (see Stoddart, 1986). The first university geography department had already been founded, in Germany, in 1874; others followed rapidly in France and (as we have seen) the UK, and later in the USA. In this period the emerging scholarly discipline was dominated by a few individuals: Friedrich Ratzel and Alfred Hettner in Germany, Paul Vidal de la Blache in France, A. J. Herbertson and Halford Mackinder in the UK, and W. M. Davis and Ellen Semple in the USA. They faced the twin and related problems of defining geographic content as a discipline, and geographic purpose in the educational system. We shall argue later that it is impossible to understand the outlines of one without heed to the constraints imposed by the other. First, though, we describe the twin components of geography as it took shape in this period. We do this not out of any intrinsic or vaguely 'antiquarian' concern for the subject's past, but rather because these focal themes arguably underpin much subsequent debate, and in large part set many of the parameters within which contemporary research is conducted. These dual edifices, then, on which geography was built, are the human–environment relationship (or the 'man–environment' relationship to use the vernacular of the day) and the regional concept.

We saw above how Halford Mackinder defined geography partly as the science that sought to examine the influence of the environment upon society. This was without doubt one of the discipline's earliest organising principles. In these formative years there was also little doubt that many geographers operated with a narrowly determinist view of this relationship, in which it was conceived as operating in one direction only: environment setting the scene for, and conditioning the evolution of, human progress. As we shall see, much of the task of early regional geography was to map these variations and to relate them to external, environmental factors. This focus was part result of, part condition for, geography's emerging status within the division of knowledge. It also owed a lot to geography's broader role within the dominant societies of the day. The environmentalist face of the discipline, then, needs to be examined at much more than surface appearance.

Two of the most frequently cited environmental determinists are Ellen Semple and Ellsworth Huntington. Semple's classic text, *Influences of Geographic Environment* (1911), sought to build on Ratzel's version of 'anthropogeography' to demonstrate the impact of different environmental factors. It began with the famous statement, 'Man is a product of the earth's surface', often quoted in support of the view that Semple was an unabashed determinist. There is much truth in this assessment. At numerous points the book stands out for its remarkable account of the ways in which the environment apparently controls human behaviour. It is replete with statements that extol the strength, power and reach of geographic environmental factors; and, as such, of course, has to be read in the context of geography's continuing efforts to establish itself as a discipline. For example,

> Geographical environment, through the persistence of its influence, acquires particular significance. Its effect is not restricted to a given historical event or epoch, but, except when temporarily met by some counteracting force, tends to make itself felt under varying guise in its succeeding history. It is the permanent element in the shifting fate of the races.
>
> (*Ibid.* p. 6)

Semple identified four classes of geographical influence: 'direct physical effects of environment similar to those exerted on plants and animals by their habitat' (p. 33 – for instance, adaptation to climate); psychical effects, reflected, for example, in literature and religion; economic and social, through the distribution of natural resources; and the movements of peoples, along with barriers to such migratory flows. She then considered a series of regional typologies, such as coast or island peoples. But it is in her description of the mountain environment that we find the following supreme account of her environmental determinist approach:

> The mountain-dweller is essentially conservative. There is little in his environment to stimulate him to change, and little reaches him from the outside world. . . . With this

conservatism of the mountaineer is generally coupled suspicion towards strangers, extreme sensitiveness to criticism, strong religious feeling and an intense love of home and family. The bitter struggle for existence makes him industrious, frugal, provident; and, when the marauding stage has been outgrown, he is peculiarly honest as a rule. . . . When the mountain-bred man comes down to the plains, he brings with him therefore certain qualities which make him a formidable competitor in the struggle for existence – the strong muscles, unjaded nerves, iron purpose and indifference to luxury bred in him by the hard conditions of his native environment.

(*Ibid.* pp. 600–1)

On the other hand it would be wrong to reduce all of Semple's work (and she wrote far more widely than this one, albeit major, text; see, for instance, Semple, 1915) to such status. Even in *Influences of Geographic Environment*, for example, there are instances where the influence of environment is admitted to be rather more complex than such superficial descriptions as that quoted above might suggest (1911, p. 31):

Anthropo-geographic problems are never simple. They must all be viewed in the long perspective of evolution and the historical past. They require allowance for the dominance of different geographic factors at different periods, and for a possible range of geographic influences as wide as the earth itself. In the investigator they call for painstaking analysis and, above all, an open mind.

Semple concluded her account with an investigation into the influence of climate, and this factor was developed further by Ellsworth Huntington in his *Civilization and Climate* (1915). Whilst Semple's environmental determinism can perhaps be accounted for by her concern for proclamation of the significance of geography as a discipline, Huntington's engagement with the influence of environment led him into even more determinist (and, ultimately, even more questionable) arguments. He set out to compare a series of maps of the earth's surface and to examine how varying distributions depended upon environment, in particular on climate. He constructed a map of the global distribution of civilisation from replies to a letter sent out soliciting the views of a range of distinguished contemporary figures, and compared this with maps of climate patterns. He concluded (1915, p. 314): 'The civilization of the world varies almost precisely as we should expect if human energy were one of the essential conditions, and if energy were in large measure dependent upon climate'. He also examined the historical record of the emergence of different civilisations and compared this to what was then known of long-term climate change, arguing that (*ibid.* p. 365) 'not only at present, but also in the past, no nation has risen to the highest grade of civilization except in regions where the climate stimulus is greatest'.

Much of Huntington's concern with climate manifested itself in a series of statements about its mediation by racial differences, and in this many of his views represented what is at least by contemporary standards straightforward racism. In one chapter, 'The white man in the tropics', he argued that 'the native races within the tropics are dull in thought and slow in action' (p. 56), explaining this in terms of the influence of climate. Elsewhere (p. 40) he attributed the lower productivity of agriculture in the southern states of the USA to the relative inefficiency of the negro alongside the white man. It is in this aspect of his work, and in subsequent statements on the significance of natural selection (see Huntington, 1924) that some of the most unsatisfactory consequences of environmental determinism were to be found.

We shall have more to say later about the significance of and limits to environmental determinism. First, however, we must introduce the second core aspect of geography as it emerged at the turn of the nineteenth and twentieth centuries: its focus on the region and the regional concept. In part this second stem of the discipline is closely related to the first, and many of its characteristic debates also rested on conceptions of the role of the environment. It remains the case, though, that there was a distinctive element that crystallised in this period around the idea of regional geography, even if there were many different interpretations placed upon the regional concept during these formative years.

Consideration of the region as a particularly geographical aspect of study owed much to the subject's origins in classical times, and its later association with voyages of exploration and discovery. Out of this past grew a concern for *identification* and *description* of particular regions of the earth's surface. Both these processes were to prove highly problematic. One of Halford Mackinder's contemporaries, for instance, A. J. Herbertson (1905), concerned himself with the process of identifying in a *systematic* fashion different regions across the globe. He defined the subject matter of geography as 'the study of phenomena from the point of view of their distribution on the surface of the earth, in natural groups, and not as isolated phenomena' (*ibid.* p. 301). He went on to argue that this entailed the definition of *natural* regions, 'definite areas of the surface of the earth considered as a whole, not the configuration alone but the complex of land, water, air, plant, animal and man, regarded in their special relationship as together constituting a definite characteristic portion of the earth's surface' (p. 301). These 'natural regions' had four dimensions – configuration (relief, altitude and structure); climate; vegetation; and population density. The latter was by far the least significant, reflecting its dependence upon the physical factors: 'A natural region should have a certain unity of configuration, climate and vegetation. The ideal boundaries are the dissociating oceans, the severing mass of mountains, or the unhospitable deserts' (p. 309). In this fashion he recognised six major types of natural region (with a variety of sub-types): polar, cool temperate, warm temperate, tropical, lofty or sub-tropical and equatorial lowland. Thus Herbertson (*ibid.* p. 309) concluded (in terms that stressed the relationship between environmental influence and regional geography):

> The recognition of natural regions gives the historian a geographical foundation for his investigations into the development of human society. By comparing the histories of the same race in different regions, or of a succession of races in the same region, it should be possible to arrive at some knowledge of the invariable effect of a type of environment on its inhabitants, and permit some estimation of the non-environmental factors in human development.[3]

Herbertson's chief concern, then, was for the identification of natural regions, using the label 'systematic' to describe the way in which his classificatory criteria ordered the space of the earth in much the same way as species ordered flora and fauna (and we should say here that the word 'systematic' is used in a very different sense in later debates). These questions, to do with the bases for and purposes of regionalisation, formed the substance for a variety of differing emphases within the early regional tradition.[4] In France, this was dominated by Paul Vidal de la Blache, whose writings focused upon the essential unity of a region – the complex inter-relations between many factors that made up an indivisible whole. In contrast to the determinism of Ratzel, he emphasised the idea of territorial unity in which human activity was an active element, not some passive recipient of environmental influence (see also Chapter 3). In *Principles of Human Geography* (published posthumously in France in 1921 and in English in 1926) he introduced (p. 10)

> an essentially geographical concept: that of environment as composite, capable of grouping and of holding together heterogeneous beings in mutual vital inter-relationships. This idea seems to be the law governing the geography of human creatures. Every region is a domain where many dissimilar beings, artificially brought together, have subsequently adapted themselves to a common existence.

Paul Vidal de la Blache's conception of a series of *pays* each with a distinctive *genre de vie* has frequently been criticised for its inappropriateness to a world beyond the agricultural age (and we shall return to this theme below). Whilst only partly correct (in his later works such as *La France de l'Est*, published in 1917, he began to address issues raised by industrialisation), such difficulties were confronted by the school of geography influenced in the USA from the 1920s to the 1940s by Carl Sauer. He, like Vidal de la Blache, reacted explicitly against environmental determinism,[5] and set out instead to investigate not only the active role of human agency but also the impact of successive stages of human occupancy upon different parts of the surface of the earth. His later writings (for instance, Sauer, 1956, p. 49) focused on the idea of humankind as guardian or steward of the earth's resources, in a sense presaging many later environmentalist debates: 'We need to understand better how man has disturbed and displaced more and more of the organic world, has become in more and more regions the ecologic dominant, and has affected the course of organic evolution'.

There were clearly, then, many different varieties of regional geography even in the early years of the twentieth century, including Herbertson's natural regions, Vidal de la Blache's *pays* and Sauer's later chorology, or geography as the description of regional associations. An early focus in some countries on environmental determinism – in the sense of environment creating the linkage of elements within regions via direct unilinear causes – gave way in later years to a more diffuse, possibilist approach in which the environment set certain limits or bounds to human existence, to which society responded in differing ways and, in turn, impacted upon the constitution of that so-called 'natural' environment (see Chapter 3). This changing conception of the regional method was important for the ways in which it set the tone of subsequent geographic debate (and there is no doubt that the influence was a lasting one: as late as 1951 Martin tried to defend determinism as a 'basic hypothesis in human geography' (p. 6) whilst, in 1966, Lewthwaite attempted to clarify the various usages of the terms environmentalism and determinism). As such it is instructive to examine not just *how* geography took the early shape it did, but also *why*.

We have already introduced the idea that geography's emerging status cannot be considered in isolation from its roles within the academic division of labour and society in general. One interpretation of the social origins of environmental determinism was that of Peet (1985b), who saw it as the geographic expression of social Darwinism, an ideology or legitimising principle drawn from nineteenth-century biology (p. 310): 'Environmental determinism, I would argue, was geography's contribution to Social Darwinist ideology, providing a naturalistic explanation of which societies were fittest in the imperial struggle for world domination'. Through the application of biological ideas such as evolution and natural selection to the study of the earth's surface and its impacts upon societies, environmental determinism (and the various versions of regional geography it underpinned) more or less consciously sustained the major world powers in their pursuit of global supremacy (see also Stoddart, 1966; Kearns, 1984). Notions such as Ratzel's *lebensraum* and the idea of states evolving like organisms, the strongest growing inevitably larger at the expense of the weakest, also underpinned colonialism and expansionism. There is no doubt that there is some truth in this Marxist interpretation of the evolution of geography (see also Chapter 2 for further comments upon geography's geo-political roles in these, and subsequent years; and note too that some scholars such as Kropotkin and Fleure saw other non-imperialist roles for geography, and did so in the context of a broader faith in a 'humanist education' – see Chapter 3). At the same time, though, it must not be forgotten that geographers themselves were, in this period as in any other, concerned not just with their role in the world, but also within the educational systems then being established. In forging those links to the rewards offered by professional status, the opportunities offered by environmental determinism and the regional method offered ready, if only roughly fashioned, pillars, around which to build their discipline and stress its difference from others. It is in the complex inter-relationships between geography, geographers and material circumstances of the day that the early focus upon these twin concerns emerged.

And these foci, we repeat, did have a deep significance. We go on to trace some of these connections in the following section, which examines the nature of the encounter between the regional tradition and the quantitative revolution.

The regional tradition, the quantitative revolution and spatial science

Whilst environmental determinism and the regional concept characterised early twentieth century geography, a range of different pressures affected the discipline after about 1945. These involved the growing tendency of geographers to adopt quantitative methods, supplementing their long-standing use of descriptive statistics with the application of inferential statistics (used to draw – subject to significance tests – inferences or conclusions of a more general nature from a limited sample or body of data). At the same time geographers began to develop the skills necessary for the analysis of large data sets through increasingly sophisticated and powerful computing facilities. In a real sense one fed upon the other: the information explosion generated vast quantities of data, which begged analysis and which, in turn, raised fresh questions, leading to a need for yet more information. In its own terms this new expertise did not amount to much; but what it did do was reconnect the discipline to the natural sciences, in the process coalescing with other concerns as geography sought to maintain, if not enhance, access to research funding and political respectability. Technical changes, in other words, enmeshed with a shifting relationship between the discipline and the broader social world, to take geography as a subject away from its earlier concern with the description of 'unique' places and into attempts at uncovering universal spatial laws governing the way in which the world worked. This intellectual shift to a discipline conceived as spatial science represented a clear change of emphasis, one that only made sense in the context of a belief in a fundamentally ordered creation, resting on universal rules of spatial organisation (and we shall have more to say about the philosophical bases of this tradition later).

There are numerous aspects to geography's engagement with quantitative methods and spatial science, but we want to argue in particular here that an important (and often forgotten) element in the rapid promulgation of the discipline's new concerns was critique both within and of the regional tradition. Scientific approaches did not emerge in a vacuum, in other words, but in a world where geography and geographers were increasingly aware of the limitations to the regional method as then practised – both in intellectual terms, and in the sense of sustaining geography as a distinctive and professional field of inquiry. This section therefore focuses first upon that debate, before charting its significance to the main elements of the quantitative revolution.

The wartime years of 1939–45 represented a significant landmark in history in a whole variety of ways, creating a moment for reflection upon past achievements and future directions. This held true in academia too. Writing in 1945, Edward Ackerman commented on geography's contribution to the war effort, and its immediate peacetime objectives, in a way that neatly highlighted both the subject's broader concern for survival in the post-war world, and the intellectual content suitable for that purpose. His starting point was the difference between regional approaches to geography – the description of particular regional assemblages – and systematic approaches – consideration of the areal distribution of different thematic or topical specialisations, such as industry or agriculture. During World War Two, he argued, it had become obvious that there were many deficiencies to a geographic education, such that it had been of only limited value to the government's war efforts. Much of this was due, he continued, to an excessive focus upon regional geography at the expense of systematic approaches:

> The regional method of research, once wartime geographic compilation and investigation was started, proved to have no more value than the past literature. Where anything more than superficial analysis was required in government work, the only possible course was one of systematic specialization. Dependable accumulations of data, and reliable interpretations of those data, were not to be had otherwise.
>
> (*Ibid.* p. 128)

Such experiences, he argued (*ibid.* p. 129),

> should cause us to take stock critically. If our literature is to be composed of anything more than a series of pleasant cultural essays, and if our graduates are to hope for anything more than teaching professions, we shall do well to consider a more specialized, or less diffuse approach.

His concerns were therefore practical ones to do with the 'mistaken view as to the simplicity of regional geography' (p. 128) – the difficulties attached to interpretation of all aspects of a given region – and the limited role of geography unless its practitioners specialised in particular topics or themes, linking together systematic and regional approaches to the subject. Only in this way, he felt, could the discipline serve a useful peacetime function, and be accepted as an appropriate scholarly discipline.[6]

Whilst passing over for the moment questions to do with the hidden value-content of Ackerman's message (in particular, his implicit acceptance that geography should serve closely defined social interests in government), it was nonetheless a revealing one in the context of the day. The distinction between different attitudes to geographic endeavour expressed a perceived difference between (and superiority of) generalising or nomothetic systematic approaches on the one hand, against essentially descriptive or idiographic regional approaches on the other. The apparent power of systematic approaches lay in their closer relationship to the natural sciences. Much of the debate over the application of quantitative methods in geography after 1945 is contained in this (ultimately false) analytical separation (and it is important to note too that the regional tradition contained more than a kernel of concern for the apparent lack of 'scientific method' in its approach).[7]

The basis for the adoption of new techniques and methods in geography after 1945 is also evident in a well-known (if often misrepresented) debate between two American geographers, Hartshorne and Schaefer. In 1939 Hartshorne codified the contemporary geographical agenda in an influential book, *The Nature of Geography*. This was the focus of an article by Schaefer (1953, p. 226), who argued, 'the methodology of geography is too complacent. Some fundamental ideas have remained unchallenged for decades though there is ample reason to doubt their power'. What he meant was that geography had become characterised by 'exceptionalism' – an idea that it was 'quite different' from the other sciences – and was thus 'methodologically unique' (p. 231), in that it sought description of unique occurrences: 'Geography according to Hartshorne is essentially idiographic. Whenever laws are discovered or applied one is no longer in the area of geography. All it contributes is facts' (p. 240). By contrast, he argued, geography should focus not on uniqueness but on similarity, by applying relevant laws in the manner of true science: 'Description, even if followed by classification, does not explain the manner in which phenomena are distributed over the world. To explain the phenomena one has described means always to recognize them as instances of laws' (p. 227).[8]

At the same time as this essentially methodological debate, another and parallel line of critique came to regard the regional concept – in the sense of regional unity – as itself no longer appropriate to changed material circumstances. This was clearly expressed by Kimble (1951), who focused on problems inherent to the definition of the regional idea. He harboured, in a famous phrase, 'suspicions that regional geographers may perhaps be trying to put boundaries that do not exist around areas that do not matter' (p. 159). This was supposedly because the world had moved on from the days of Paul Vidal de la Blache, to a situation where regional identity was far harder to define largely because it had been swamped by broader national and international processes (Kimble, 1951, p. 159):

The airplane has made the whole world a neighbourhood, and made it just about as private. The effects of the new mobility are legion, and none more significant than the accelerated tempo of cultural diffusion . . . of no single part of the earth can it now be said that the pattern of living is a straightforward product of evolution brought about by local and internal forces.

Whilst the regional concept might have been applicable to agriculturally dominated societies, he argued, where a harmony between society and nature seemed to exist down even to distinctive building styles (due to the use of local building materials), the world had moved to a different age:

> We must, then, face the fact that the world is changing. . . . Whatever the patterns of the new age may be, we can be sure that there will be no independent discrete units within it – no 'worlds within worlds'. There will be no neatly demarcated 'regions' where geographers (or economists or sociologists, for that matter) can study a 'fossil' community. Man's region is now the world.
>
> *(Ibid.* pp. 172–3)

Kimble's contribution to the debate over the regional tradition neatly prefigured many subsequent geographic concerns, to do with the relationships between different levels of analysis and planes of inquiry. These are encapsulated in dialogues over the significance of locality and of agency and structure, for instance, both of which are described later in this chapter and at length later in this book. Kimble's interpretation rested on a particular reading of the regional unity concept, which emphasised local elements as opposed to the role of different regions within a larger space; and, as we have seen, this downplayed both the variety within the regional tradition and the concerns of some classical regional geographers for broader social processes. But in the sense of announcing the severance of regional geographic study from the physical influence of landform and climate, the 'organic' unity of some early versions, he did touch on highly important themes that went to the very core of the regional tradition.

That they were significant can perhaps be gauged by the evident concern on the part of some regional geographers of the time to re-state and rework the primacy of the regional method. For example, James (1952, p. 199) defined the region as a 'geographic generalization' (borrowing on a term from the scientific language of the day), isolated only in relation to a particular problem. He went on to stress the search for process in regional geography in and through its quest for significant associations between phenomena in particular places; from description came explanation. Whittlesey (1954, p. 30) similarly defined the region as a 'device for selecting and studying areal groupings of the complex phenomena found on the earth'. He concluded (p. 65), 'the need for regional study is more urgent than ever'.

The stridency of such defences points to the increasing assault of spatial scientific approaches upon mainstream geographic thought. Several turned their attention specifically to the 'new' geography and examined its apparent limitations. For instance, Stamp (1957) reaffirmed his belief that the primary concern of geography should be the mapping of data, reflecting even on the choice of colours to be used. He went on (p. 2):

> I am, however, a little alarmed by the view that the geographer must add to his training a considerable knowledge of statistics and statistical method, of theoretical economics, and of modern sociology. Sufficient perhaps to appreciate what his

colleagues are doing, so that team work may be based on mutual appreciation, seems to me the right attitude.

And Spate (1960, p. 392) also reflected on where the tide of quantification was leading, in his call for balance between 'quantity' (i.e. quantification) and 'quality' (i.e. interpretation): 'The crux, as it seems to me, is this: quantification is in the end essentially classificatory rather than truly interpretative, though it is often an essential tool towards interpretation'.

By the early 1960s, though, advocates of change such as Burton (1963, pp. 152, 155) could assert, perhaps a little airily, that

> The quantitative revolution in geography began in the late 1940s or early 1950s; it reached its culmination in the period from 1957 to 1960, and is now over. . . . The erstwhile revolutionaries are now part of the geographic 'establishment' and their work is an accepted and highly valued part of the field.

For Burton, dissatisfaction with idiographic geography lay at the heart of the quantitative revolution, and the development of a theoretical, model-building geography was likely to be its major consequence. It was increasingly apparent that the process of quantification was in itself insufficient to guarantee such theorisation:

> The need to develop theory precedes the quantitative revolution, but quantification adds point to the need, and offers a technique whereby theory may be developed and improved. It is not certain that the early quantifiers were consciously motivated to develop theory, but it is now clear to geographers that quantification is inextricably intertwined with theory.
>
> (*Ibid*. p. 157)

In this way the adoption of quantitative methods set the scene for a growing engagement with other issues of method, to do with the proper and rigorous pursuit of intellectual inquiry within a scientific framework. These involved adoption of the view that there existed spatial laws or rules, which (if only geographers could uncover them) would prove to be at the root of all human existence. The scientific methodological tilt of quantification, then, led to a growing engagement within geography over the notion of spatial science, concerned with formulating predictive laws of spatial behaviour. As William Bunge put it in the introduction to the first edition of his *Theoretical Geography* in 1962 (p. x): 'The basic approach to geography is to assume that geography is a strict science and then to proceed to examine the substantive results of such an assumption'. By 1966, in an introduction to the second edition (pp. xv–xvi), he had enlarged upon such remarks as follows:

> The earth is not randomly arranged. Locations of cities, rivers, mountains, political units are not scattered around helter-skelter in whimsical disarray. There exists a great deal of spatial order, of sense, on our maps and globes. . . . This period sees geography emerging as the science of locations, seeking to predict locations where before there was contentment with simply describing.

Similar concerns underpinned Haggett's influential *Locational Analysis in Human Geography* (published in 1965). For as Golledge and Amadeo (1968,

p. 760) argued, the quest for laws and theories was a 'logical outcome of the continued search for scientific explanation of the spatial relations of phenomena'. From quantification, then, there grew increased concern for respectability (in the eyes and language of the natural sciences at least) via the application of scientific method, leading to the notion of geography as a spatial science concerned with modelling and predicting human spatial behaviour.

Much of this rested on a series of more-or-less implicit assumptions that the world was somehow governed by natural and unswerving laws (that it was an ordered creation), which reconnected geography with the various philosophical traditions of positivism (see, for instance, Gregory 1978, Part I). Whilst spatial science – with its search for laws in empirical regularities – was undoubtedly underpinned by a positivist philosophy, very few spatial scientific geographers spent much of their energy upon such issues, preferring to concentrate instead on questions of method. M. R. Hill (1981) neatly described positivism as the 'hidden philosophy' in geography; even David Harvey's *Explanation in Geography* (1969b), the foremost scientific methodological text of the day, said virtually nothing about positivism as philosophy. Yet the search for new forms of methodology led ultimately, for some, to (at least temporarily) blinding truth. As Harvey put it in the preface to *Explanation in Geography* (p. vi), the integration of quantification and scientific principles within geography 'opened up a whole new world of thought in which we were not afraid to think theoretically and analytically'. Yet only four years later in his *Social Justice and the City* (1973), Harvey was to abandon much of this new-found explanatory power and to move instead through liberal to Marxist historical materialist analysis as the route towards intellectual advance (p. 17):

> The emergence of Marx's analysis as a guide to enquiry (by which token I suppose I am likely to be characterised as a 'Marxist' of sorts) requires some further comment. I do not turn to it out of some *a priori* sense of its inherent superiority (although I find myself naturally in tune with its general presuppositions of and commitment to change), but because I can find no other way of accomplishing what I set out to do or of understanding what has to be understood.

Such a radical shift from one so heavily implicated in the spatial-scientific revolution was deeply significant, though Harvey was far from alone. Others, increasingly dissatisfied with the restrictions of logical positivism, made a similar break. They were joined by a growing range of scholars, too diverse in their interests to merit the term 'group', who sought alternative ways of interpreting and understanding the world. These individuals made use of the intellectual space and comparative freedom created within geography by earlier developments. In this very real sense, then, the spatial-scientific revolution in geography set the scene for rapid changes in subsequent years. And in the following section we turn to the emergent critique of spatial science and introduce the growing diversity of alternative approaches that form the essential core of this book.

The critique of spatial science and the emergence of alternative approaches in the 1970s and 1980s

During the later years of the 1960s and beyond, a number of objections were raised, generally with increasing sophistication, to the model of geographic intellectual inquiry as represented by spatial-scientific formulations. These led in various directions, many of which are the subject of the remainder of this book. First, however, we must consider the substance of these criticisms and examine some of the responses to them; and briefly mention the continuing regional thread that remained in geographic practice despite the debates of preceding years. Once these tasks have been completed, we turn to a brief introduction to the distinctive substance of alternative approaches in the 1970s and 1980s.

There are, then, a number of ways in which spatial science and its positivist underpinnings can be criticised. In the first instance it arguably creates a false sense of objectivity by artificially separating observer from observed, denying the existence of strong correspondence links, and asserting value-neutrality on the part of the observer. Similarly, its use of mathematical or geometric languages filters out social or ethical questions, allowing no room for consideration of values or how society could be organised as opposed to how it actually is organised. More generally, spatial science suffered from a failure to 'see beyond the map', for it did not acknowledge two crucial aspects of spatial patterns and processes. They are, first, intimately bound up with the working of deeper economic, social and political structures that condition and constrain the paths of human existence (an early concern of the Marxist critique – see below); and, second, they are reflected in and are reflections of the perceptions, intentions and actions of human beings as conscious agents (the starting point for an alternative humanist approach – again, see below).

Many of these weaknesses and problems have subsequently been recognised by proponents of spatial science, who have taken on board a more conscious recognition and evaluation of the philosophical implications of scientific methodologies (exemplified in Popper, 1959; Kuhn, 1970; Lakatos, 1970). For instance, Hay (1985, p. 137) admitted that 'the enthusiastic practitioners and protagonists were often unaware of the problems inherent in the scientific approach, and did not clearly identify the additional problems posed by its geographical use'. And Cox (1989, p. 208) similarly cautioned against excessive dependence upon both models for their own sake and large data sets as the route to scientific explanation, worrying that 'too much research in the Brave New World will be guided by whatever data happen to be readily and copiously available. The danger will be . . . that preoccupation with data will limit creativity'.

In a similar vein to Hay, Bennett (1985, p. 213) accepted that for 'most geographers . . . the spatial geometric base of spatial science has little major utility'. He commented (p. 216) upon a clear re-appraisal of the inferential statistics that had been at the heart of early quantitative work as follows (for an earlier cautionary statement, see Gould, 1970): 'a considerable internal debate within the corpus of quantitative geographers has led to the rejection of most of the earlier text-book presentations of quantitative methods, and also the terminology and methods of statistical inference as the dominant mode of quantitative analysis'. In particular, he continued, geographical data sets were mediated by the structure of data collection – the areal units used to generate and order information. This 'modifiable areal unit problem' (see also Openshaw, 1984) breached the strict scientific assumptions of independence in observation and sampling; but there were, argued Bennett, adequate alternative quantitative methods more appropriate to the geographical situation. He responded to the critique of quantitative geography as positivism, arguing that to equate the two was not sustainable; concluding (p. 223) that quantitative methodology should instead be regarded as 'a tool developed for use within the value position or the research question in hand. In itself it is neither a ready-made philosophy nor a value-free method of research'.

Whilst many accounts of the diversity of approach that emerged in the wake of the quantitative and scientific revolutions often treat the latter as if they had ground to a halt in the late 1960s, we would stress here that there has been a continuing and productive engagement with issues of methodology (and, increasingly, of philosophy) within this strand of the geographic web. Equally, it should not be assumed that the regional tradition was somehow ground out of existence in this era. On the contrary, a host of regional studies and courses continued to thrive, some as if nothing had happened, others with greater regard to the potentially enabling qualities of the new environment for geographic research. Such concerns exercised Paterson (1974), who presented 'a practising artisan's commentary on the current state of his craft' (p. 4). He recognised that there were many problems in writing regional geography in the contemporary world, many of which we have addressed above; but also argued that the possibility was there for fresh progress in the context of other changes:

> Given the development of geography in other directions – the growth of theoretical and quantitative concepts, and the huge increases in statistical resources and data processing facilities – it would seem that the impact on regional geography ought to be liberating . . . . The regional study can concentrate on being precisely that . . . . The way is open for regional studies which are less bound by old formulae; less obliged to tell all about the region; more experimental and, in a proper sense of the word, more imaginative than in the past.
>
> (*Ibid.* p. 23)

Whilst Paterson's version was only one of many possible alternative directions for a continuing regional geography, it is significant here for its reassertion of the regional method in the specific context of the 1970s.

Clearly, however, there were other, more radical departures taking place in this period: and it is to these our account now turns. As we have seen, there were at least two significant and broadly discernible avenues of departure for critiques of spatial science: one focusing upon the constraining effect of social, economic and political structures, and another more concerned with the mediating influence on spatial patterns and processes of the creativity and independence of thought of human agents. Whilst such an analytical separation was subsequently to become increasingly blurred, as the concerns of these 'schools' intersected, in the account that follows we trace first (in Chapter 2) the development of broadly Marxist approaches to human geography during the 1970s and 1980s, followed (in Chapter 3) by the formulation of humanist approaches. The developing encounter and dialogue between these led to an increasing awareness of the complex nature of the social world and was intimately bound up with a greater openness to the literature and debates of the wider social sciences (as well as to the ideas of philosophy and the humanities). One particularly pressing problem was the relationship between agency and structure (and here we suggest that there were more than just echoes from the geographical past, to do with the influence of environment on human action). In Chapter 4 we consider one suggested route to resolution of the 'problem' of agency and structure, the structurationist approach in which agents and structures are conceived of as mutually and immediately inter-related. Then in Chapter 5 we consider a (partly linked) approach that rests upon wholly different philosophical foundations, those of realism. Finally in Chapter 6 we evaluate the recently growing engagement with postmodernism by some human geographers, and tentatively identify possible future areas of debate. Below, these approaches are reviewed briefly in turn.

An early concern of Marxist approaches to human geography, then, was the evident disregard for deeper structural conditions of social existence contained within the spatial scientific agenda. In analysing patterns of urban inequality and deprivation, for example, Marxist geographers pointed to the impact of factors such as local and national government policies, and the activities of financial institutions (see, for example, Harvey, 1973). This focus on the broader determinants (as they seemed at the time) of spatial patterns coalesced with a growing tide of critique of the relationship between geographic research and the broader world; for widespread political radicalism in the late 1960s led some geographers to question earlier assumptions about the purpose of geography and its connections with government objectives. Not just the content, but also the intent of geography became a matter of some debate (the latter especially not for the first time, but certainly in the first sustained exercise for many years). In this way Marxist approaches developed an early emphasis on grounding geographic research in the material conditions of existence, in the historical-materialist mode of analysis. And this relationship between Marxist approaches and the broader social world is used as an ordering device in our Chapter 2, which identifies and describes four phases in the development of Marxist human geography.

The first relates to the emergent years of the late 1960s and early 1970s when a prime concern was the integration of a specifically Marxist critique from within a broad band of more widely radical writings. In this process the role of a few individuals – many of them one-time figure-heads of the quantitative revolution, such as David Harvey – was crucial in giving shape, substance and a fair degree of legitimacy to the new (and it was that, to all intents and purposes, despite the relatively long social scientific pedigree of Marxist analysis) strand to geography. Once established to an (admittedly limited) degree, Marxist geographers engaged upon a second phase in the mid to late 1970s, characterised by a search for mature reflections. They worked on topics to do with issues such as development and imperialism in the Third World, uneven regional development and urban inequality (see Peet, 1978). In these agendas, it increasingly became evident that the early focus on structural influences (and the emphasis placed upon structuralist accounts and versions of Marxism), whilst an adequate starting point, was also capable of hindering further progress (see Duncan and Ley, 1982). Thus in a third phase characterised externally by continuing recession and influenced in the academic environment by retrenchment and regrouping, Marxist geographic approaches gradually drew more heavily upon other concerns and versions of social theory, linking structural influences analytically to the workings of the world. In the process, though, this broadening of concern led to a fourth, most recent phase in the late 1980s, characterised by a dialogue with alternative analysis. In the process

Marxist claims as to the significance of class relations were subjected to reformulation and re-evaluation. Established within the geographic environment, so to speak (see, especially, Harvey, 1982; Massey, 1984), Marxist approaches faced a difficult challenge of maintaining relevance to both intellectual and practical issues of the day.

In Chapter 3 we turn our attention to the development of humanistic approaches. In a parallel fashion to our treatment of Marxism in human geography, this account takes a loosely chronological view of the (re)discovery of humanist principles on which to ground geographic research. We say 're'-discovery because, unlike the adoption of Marxist theory, there is a longer history of geographic engagement with such humanist concerns, from the 'scientific humanism' of the Renaissance through to the treatment of human agency in some possibilist versions of the regional tradition. Much of this disappeared under the weight of spatial science in the 1950s and 1960s, only to resurface alongside the Marxist critique. In part this was a result of impulses from within the spatial-scientific tradition, to do with the application of stochastic principles and an interest in human behaviour, which led to the early versions of (still undoubtedly and unashamedly positivist) behavioural geographies. However, the evident failure of spatial science to take seriously the complexity of human beings as creative individuals led to a growing oppositional critique founded on more strictly humanist principles. Whilst Marxist versions reacted against social inequality and the uneven distribution of power in society, humanist geographers focused initially on the role of human agency, and in particular on the way in which the researcher's subjectivities mediated the research process.

We argue in this chapter, then, that the adoption and subsequent reformulation of humanist approaches during the 1970s and 1980s has been a more fragmented process than the evolution of Marxist thought in geography (in spite of more recent events within the Marxist tradition), in part because the humanist label covers a diversity of different and sometimes incompatible intellectual positions. We also argue that, whilst early humanistic geography (during the 1970s) took up the traditional humanist engagement with the personality of the researcher (see, for instance, Relph, 1970; Tuan, 1976; Ley and Samuels, 1978a), more recent versions of what might be characterised as a 'geographical humanism' have reversed these priorities and focused more upon the humanity of the researched, of the individuals who are the focus of study (and particularly important in this respect has been the ongoing work of Ley). Whilst all of this is not straightforwardly compatible with the more philosophical turn to humanist thought given by phenomenology and existentialism, it represents a challenging departure for human geography, particularly as humanistic approaches begin to encounter and engage with the historical materialist version of Marxist geography (see Kobayashi and Mackenzie, 1989).

This dialogue between the twin focal elements of agency and structure forms the point of departure for our two following chapters. In Chapter 4 we introduce and outline Anthony Giddens's version of the theory of structuration as one means of resolving the tension between agency and structure, and as a possible touchstone of geographic research. Structuration theories (of which the prolific writings of Giddens represent only one, albeit significant, kind) have four common concerns (see Thrift, 1983). They are antifunctional; that is to say they seek to avoid the functionalist explanation of some structuralist versions of social science, wherein something exists only because it is functional, and is functional only because it is seen to exist. They recognise the duality of structure; in other words, structures are constructed out of the efforts of human agency and are at the same time constitutive of that social action. They seek a theory of action; and finally, they emphasise the significance of time and space in the constitution of society. Giddens's account of structuration theory is important not just for social theorists but also to human geography, for he has engaged with and criticised the notations of time-geography, that version of geographic theorisation concerned with the significance of paths and projects of activity constrained in time and space (see, for instance, Pred, 1977; Hägerstrand, 1982). The ideas of Giddens have been introduced to the geographic literature by the writings of Pred, with his notions of place as 'historically contingent process' (see Pred, 1984a) and, in particular, Gregory (see, for instance, Gregory, 1989b; 1990b). We conclude on a cautionary note though, in showing that – whilst structuration represents an attempt at constructing a grand theory for the understanding of society – there are great problems in transferring its ideas directly into practical research. Structuration theory, in other words, is perhaps best regarded as a series of warnings about how not to approach human geography rather than as a blueprint for how to do so.

The significance of realist approaches to human geography is the subject of Chapter 5. Realism sets out to provide a wholly different foundation to positivism (and, as we shall see, one measure of its 'success' is in the extent to which it achieves exactly that), on the grounds that there are many flaws in positivist assumptions. Knowledge can come from participation, not just observation; language is not the only form of communication; knowledge is not a finished product; and science is not necessarily the highest form of knowledge. Whilst positivism is grounded in the search for regularity, realism seeks causality primarily in the nature of objects themselves and only secondarily in their interactions with each other. There are many different kinds of realist philosophy (see, for instance, Bhaskar, 1975, 1979; Harré, 1979, 1986), but realism has been introduced to a geographical audience principally by the work of Sayer (see, for instance, Sayer, 1984), who focused on criticisms of positivism in human geography. Positivism, he argued, conceptualised constant mechanisms for causing regularities, and constant circumstances in which those mechanisms operate: twin conditions suited to the closed system more common to the natural sciences but untenable in the open systems of social science. He went on to investigate the nature of relations between objects, distinguishing those that are necessary from those that are contingent; and distinguishing abstract concepts that isolate in thought partial or one-sided aspects of an object, which is itself concrete. He used these distinctions as the basis for an account of methodological research issues, arguing that geography required theoretically informed concrete research; he went on to specify appropriate research strategies for these goals.

Our account considers these aspects of realist philosophy as they have been put into practice in different research contexts: in the analysis of patriarchy, for instance (see Foord and Gregson, 1986) and in regard to the concept of locality (Cox and Mair, 1989). It concludes with three particularly significant issues in the evaluation of realism. In the first instance realist philosophy is essentially eclectic, lacking identity and drawing heavily upon other foundations. Second, it can be argued that it lacks the right methodological tools to achieve the tasks it sets itself. And finally, realism proposes an extra-theoretical reality that is approximated to in the research process, and that is also the ultimate judge of theory; in this sense, therefore, it is not that different from positivist conceptions. We conclude in this chapter that the jury is still out on realism on two counts: whether or not it is a useful and fruitful basis for further research, and whether or not it represents little more than a pathway to the approaches of postmodernism.

In the final chapter, then, we turn our attention to geography's growing engagement with the issues raised by the postmodernist critique. Postmodernism is, for many, infuriatingly difficult to define: its celebration of difference and disavowal of grand or totalising theory makes it a particularly diffuse conceptual notion with many varied meanings and emphases. Here, we start with the distinction between postmodernism as object of study – as 'epoch' in its own right – and as an attitude towards knowledge. Geographical research has tended to favour the former, developing from the postmodernist architectural association to changing conceptualisations of the built environment and, more problematically, to the way in which economic production is organised and then bound into the workings of social, political and cultural realms (see, for instance, Lash and Urry, 1987; Harvey, 1989b; Soja, 1989). Postmodernism as attitude, however, tackles head on the philosophical assumption shared by so many intellectual positions that there *is* a degree of order in the world, a rationality to creation that structures and governs the way in which it works. In this sense, then, postmodernism is based upon incredulity towards 'metanarratives' – to grand plots that encompass the sweep of human existence – in part because it supposes such metanarratives to be insensitive to the differences between different peoples and places. In this way in particular, postmodernism is a far from uncontroversial notion, capable of many different interpretations (and the discerning reader will find in our overall account and especially the Epilogue that this also applies to the present authors). There is undoubtedly a reactionary or neo-conservative tint to much postmodernist thinking: an acceptance of the status quo and a disregard of reform. What often characterises reactions to postmodernist attitudes is the extent to which it is possible (indeed necessary) to envisage, let alone formulate, a more radical version based (in geography at least) upon sensitivity to place and the differences between regions and peoples. Some of these debates are taken further in the Epilogue to this book. To extend our earlier analogy, if the jury is out on realism, then the case has not yet come to court for postmodernism, but evidence is being prepared, although postmodernists would not recognise the jury, claiming that there is no over-arching court of appeal to judge the relative merits of different intellectual positions.

In this introductory sketch of the diversity of approaches found in contemporary human geography, then, we have emphasised the evolution and significance of different ideas and conceptual frameworks, briefly illustrating some of the main points that are the focus of subsequent chapters. In this there has been insufficient time and space to dwell at much length upon the broader conditions in which such developments took place, in contrast to our consideration of earlier stages in the history of the discipline. We cannot wholly compensate for this here, for to do so would entail writing another, different kind of book.

What we can do, however, is not so much reiterate the significance of geography's relationship to the broader material world (although note that this is a central theme of Chapter 2) as illustrate, briefly, the importance of geography as a lived entity, occupied and created by individuals who are themselves part of the duality of structure via their relationship to centres of formal geographic education. In the following section, therefore, we go on to consider the significance of disciplinary boundaries and academic career paths.

Geography, disciplinary boundaries and academic career paths

The history of geography is of course not just bound up with the sterile development of ideas and concepts in some kind of personal and political vacuum. In addition to the pressures and constraints imposed and the opportunities offered by changing material circumstances in the wider world, it is also important not to forget the concrete history of the discipline of geography and the significance of individual career paths. The former, of course, is much more than the sum of the latter, for the structure of geographic education – from school to university – reflects a far broader set of decisions and historical traditions to do with the content and objectives of what has come to be known as geography, and its intellectual and practical significance. Geography as a discipline, then, is not just an object but a lived subject, peopled by individuals who play their (in the main, relatively small) part in its construction and reformation, retreat or advance. Such concerns run through any debate to do with the nature of geographic thought and practice. In this sense the content of geography is a reflection of two sets of interacting assumptions, to do with the boundaries drawn around an academic division of labour delineating 'Geography'; and the role of individual and collective academics in shaping, fashioning and occasionally overturning those boundaries, whilst simultaneously contributing to an ongoing process of change *within* those same constructions. Whilst the academic division of labour is so often self-evidently artificial, it is none the less both real and significant.

The most immediately obvious manifestation of this is the existence of university-level departments of geography; yet it is all too easy to take this for granted. That geography exists as a discipline is due in part at least to the intellectual battles waged within various seats of higher education to establish it as such. Perhaps the sharpest reminder of this comes from those places where these skirmishes have been unsuccessful. For instance, N. Smith (1987a) investigated what he described as 'academic war' over the field of geography in the elimination of the subject as a separate department at Harvard University in 1948, after a long-standing dispute with the department of geology over the appropriate division of intellectual labour. The then Professor of Geology accounted for the demise of geography as follows: 'Harvard can't hope to have strong departments in everything' (*ibid.* p. 159), but as Smith (p. 155) also noted: 'The blow was all the more severe because the decision to eliminate geography at one of America's leading universities was justified at the time by the suggestion that geography may not be an appropriate university subject'.

At Harvard, then, geography lost a round in the academic infight. Such concerns have an echo today at a time of financial stringency in universities engendered by government expenditure cutbacks, and on more than one occasion have led geographers to voice uncertainty over the future status of the subject. In his presidential address to the Association of American Geographers (AAG), Abler (1987, p. 518) expressed this as follows: 'We are a small discipline. To be small these days is to be vulnerable, even if the quality of our work is high'. He went on (*ibid.*) to sound a warning and to call for greater common purpose within the discipline in a bid to protect its status: 'Geography's continual splitting into smaller and smaller clusters has become hazardous to our collective health'. In similar fashion, Guelke (1989b) argued that geography was insecure at university level (in the USA in particular), having lost the intellectual coherence it once had on the basis of environmental determinism. Geography, he insisted (p. 125), had become characterised by excess internal diversity so that 'Some geographers would be hard pressed to identify the distinctly geographical component of their work, and many have more in common with scholars in cognate disciplines than those of their own'. In words reminiscent of Abler's, he concluded (p. 129), 'In the long term the discipline's survival is dependent on geographers developing a stronger sense of disciplinary purpose'. Such views on the necessity of disciplinary survival at all costs were not of course shared by all. One of the thrusts of the Marxist critique of geography in the 1970s and 1980s was the need to transcend arbitrary academic conventions (see Chapter 2). Yet it remains the case that the content of geography is largely shaped by historical traditions on the appropriate division of academic labour.

At the same time, though, it is important not to lose sight of the significance of individual career paths within these frameworks. One way of measuring this is through an analysis of printed output such as books and articles. For example, Stoddart (1967) charted the growth and history of geography in terms of the numbers of geographers at various institutions and their respective 'production'. Citation analysis is also often employed to assess the extent to which particular contributions or schools of thought are referenced in the academic literature (for an example, see Whitehand, 1990). In these and similar ways the role of key individuals within an academic structure can be evaluated and assessed.

An alternative perspective on this relationship between academic and academia – between geographer and geography – is to focus on the significance of new ideas as a means not just of overturning old orthodoxies but also of overthrowing old establishments. One interpretation of the quantitative revolution rests on just such an understanding: that of Taylor (1976): He saw the process of quantification as an attempt to displace an existing hierarchy within the discipline by a new generation of geographers. Like any other revolution, this was not accomplished without hostile opposition from the 'traditionalist' old guard (and note also that alternative interpretations of the quantitative revolution are possible: see, for instance, Stoddart, 1986).

It is not just in such radical breaks, though, that geographers create geography; nor should a focus on key individuals detract from the role of other scholars. One of our central concerns in this volume is to highlight what might be called the 'personality and personalities' of geographical endeavour; the impact of many, many individuals through their career paths and their active (re)creation of geography through teaching and research. Such a focus was explored in systematic and detailed fashion by Buttimer (1983) in her edited collection of autobiographical essays and round-table discussions on the development and nature of geography. She presented (p. 2) 'stories about the nature of geography – individuals telling about their own experiences and the context in which their life journeys unfolded'. We are unable to use such an approach here for reasons of space, time and resources, but what we do want to do is to convey something of that *personal* element of geographical experience.

Accordingly, in the remainder of this section we report, briefly, a few individual geographers' reflections upon the nature of change in human geography since the late 1960s. These are drawn from two sources: first, Macmillan's (1989) collection, *Remodelling Geography*, commenting on the two decades since publication of Chorley and Haggett's *Models in Geography* in 1967; and, second, a brief, highly 'unscientific' questionnaire we despatched to a group of geographers in 1989 asking for their personal impressions about theory in human geography.

Remodelling Geography was a conscious attempt to consider progress in modelling since the late 1960s; as Macmillan put it in the Preface (p. ix), it aimed 'to counteract the notion . . . that modelling is something that people did in the sixties and seventies but have now largely abandoned'. In his own chapter (*ibid.* p. 107) he made a stronger claim: 'Quantitative theorising in human geography still has much to offer. Indeed, with a greater clarity of purpose and a changing portfolio of subjects, modelling ought to command a central place in the discipline for another twenty years at least'. Yet the volume also contained contributions from others who had little time for such views; individuals whose life-paths had drawn them away from the theoretical position of positivist spatial science, into different pastures. David Harvey, for instance, reflected on his shift 'from models to Marx', arguing that 'those who have stuck with modelling since those heady days have largely been able to do so, I suspect, by restricting the nature of the questions they ask' (*ibid.* p. 212). His own development from *Explanation in Human Geography* (1969b) onwards was described in terms of the kinds of question he addressed: 'perhaps it was because I was trying to theorise about the historical geography of capitalism that I broke with most of my colleagues and sought for a different though equally rigorous path for building theoretical understandings' (Harvey, 1989a, p. 213). The route to explanation was significant, for Harvey (*ibid.*) expressed surprise that geography had not drawn upon Marxist approaches earlier: 'It seems extraordinary in retrospect that Marx was so lacking in Geography . . . . I am at a loss to explain the absence of any explicit Marxist tradition in 'Western' geography. But in 1967, Marx and Marxism were scarcely to be found in geography'. One result of this lack of a Marxist tradition within geography was (*ibid.* p. 215) 'a willingness to experiment and make mistakes. It also produced an atmosphere of vigorous confrontation and excitement'. Within this climate of turbulent endeavour, many dramatic shifts in the conception and application of geographic theory were initiated.

Like Harvey, Denis Cosgrove also rejected spatial science, but for a different set of reasons and with differing results. In the late 1960s Cosgrove was on the receiving end of a geographical education at Oxford University. He described (1989c, p. 230) the limited impact of the quantitative revolution there as follows:

I was aware, as were my contemporaries, of a deep divergence between the discipline of geography as presented in the Oxford School and the subject as it was being promoted elsewhere – in the quantitative schools of North America such as Pennsylvania State and Northwestern for example, and Cambridge and Bristol here in England. At Oxford we were following an undergraduate geography programme still essentially based on the ideas of the founders of British geography, A J Herbertson and Halford Mackinder. It was constructed on the study of regions coordinated into a hierarchy of scales from the global, defined by climate and topography, to a comparative national study of France and Great Britain, and finally, as our own individual dissertation, a geographical description of an area no more than 150 square miles in extent. This description had to encompass the physical and human character of the area. Oxford was defensively proud of its regional tradition.

He went on (*ibid.*):

It would be an exaggeration to claim that this course of study was intellectually stimulating, or that we as students were entirely contented with Oxford regionalism. We were aware that away in the Fens or over the Avon Gorge [at Cambridge and Bristol] a very different conception of geography was being presented to our peers. 'Geographical models' was the phrase which captured this new geography.

Once exposed to the growing presence of spatial science, though, Cosgrove (*ibid.* pp. 232, 235, emphasis added) found this equally unsatisfactory:

it was above all a distrust of the claim attached to theoretical and structural modelling in the 1960s, that the truth of geographical relations could be captured in the cold language of figures, that formed the core of my rejection of this approach. . . . I did not want the simplification of reality which [models] offered. Indeed, that claimed simplicity appeared increasingly bogus. . . . Model building . . . seemed to proclaim the objectivity of geometrical and mathematical models as themselves the guarantors of truth in representing geographical reality. It was here that a *geographical humanism*, seeking to draw upon direct experience rather than abstract reflection, opened a fundamental critique of model-building and quantification.

There are a number of significant features to remark upon in these two contrasting accounts. For Harvey, Marxism; for Cosgrove, humanism: different responses to a dissatisfaction with spatial science led in different directions, both highly productive for future development in geographic theory. Equally, the institutional context is important: the poles of advance in scientific method in the late 1960s at Cambridge and Bristol contrasting strongly with the established orthodoxy at Oxford and elsewhere. (It is interesting to reflect that today Macmillan – editor of *Remodelling Geography* – teaches at Oxford.) In North America, too, spatial science was strongest at a few institutions. As Robert Lake put it to us, for instance:

The University of Chicago, during the years I was in residence (1970–74) was steeped in theory – of urban spatial structure, theories of urban community, theories of positivist analysis of cities. Geography at Chicago was dominated by two influences: (1) The Chicago school of human ecology – and its extensions to social area analysis, factorial ecology, etc; and (2) The quantitative revolution, personified by Brian Berry – probably *the* dominant force in the department. All of this was highly theoretical and analytical, and it was what graduate students were taught at the time. It was also highly constrained and limited. Some developments and departures were tolerated – for instance, into behaviouralism with its strong analytical and psychological basis.

But Marxism was *never* mentioned. I learned about Marxist theory in urban geography in my own time, and only after I left Chicago and discovered that there was another world out there that no one had told me about. So, yes – 'theory' was highly time and place specific – filtered through the viewpoints of a few individuals.

Such a socialisation into geography contrasted with that of Trevor Barnes:

> By the time I entered geography in the mid-1970s positivism was well on the wane; in fact at the University College, London geography department no one who was serious seriously believed in it, and so the lectures on positivism I did get were always half-hearted. After all, we were post-*Social Justice and the City* kids, and we all knew what Harvey thought about quantitative methods. In general, the most interesting stuff going on for me in geography was clearly on the other side of the Atlantic at Clark and Johns Hopkins. It was always a minor event when the latest issue of *Antipode* arrived at the geography library. Although I attended a class taught by David Lowenthal, who used Tuan's *Topophilia* as the class text, humanistic geography never seemed to cut it compared to radical geography.

The path of diffusion of new ideas and approaches – especially across the Atlantic – was often an uneasy one. Ian Simmons reflected on a year spent at York University, Toronto, in 1972–3, away from Durham University, as follows:

> Now Toronto was different: a young unit with a name to make, an enterprising chairman and at least two products of the mainstream quantitative revolution traditions. I listened to them at seminars and was quite impressed by the foreignness (to me) of their lingo and their concerns; I remember feeling that here was a great area that I was ill-equipped to deal with in for instance undergraduate tutorials once I got back to the UK. But the great star of the year was the presence as Visiting Professor of Bill Bunge. I thought he was very eccentric; now I hope I'd be more tolerant but he had the annoying habit of not answering questions and retreating behind a screen of getting annoyed. He was of course into the Radical 'Expedition' mode of human geography beside which everything else lacked validity: he was tunnel-vision personified but got a largish following among the graduate students, though less so in the Faculty. Bill was probably my first experience of a really committed human geographer, ie obviously and publicly committed 24 hours a day. Quite frightening in some ways, of course.

The over-riding impression, though, from responses to the questions we asked, was that theory as such in human geography was not widely taught in the 1970s. Jacquie Burgess typified many replies: 'How was I taught theory in human geography? The short answer is that I wasn't!' She concluded with a reflection that neatly encapsulates the extent of theoretical advance within human geography in the intervening years, and the gulf of communication that all too often still remains to be bridged:

> In more general terms, it seems to me that theory in human geography is becoming more confident of itself – especially, I suppose, because Giddens chose to emphasise space and time which we could all recognise as 'geographical' concepts. There is much less sense now of geographers simply chasing around trying to find the most recent (or not so recent) -ism, or -ology and simply plug it in. Harvey is also terrifically important for us, I think, because he has a reputation which goes well beyond the discipline and he does write so well. I am more concerned about the 'obscurity' of many overtly theoretical

papers. Why are they so difficult to read? Does it denote a lack of confidence still lurking beneath the surface – or are the contemporary theorists content to communicate only to the chosen few – a new kind of theoretical freemasonry?

It is in a rejection of that 'freemasonry', and an attempt to bridge the communication gap, that this and the following chapters are grounded.

Concluding comments

In this chapter, then, we have sketched out (all too briefly) some of the key moments of the geographic discipline as it emerged at the end of the nineteenth century and at the start of the twentieth century. We focused in particular upon the idea of environmental determinism and the significance of the regional tradition. Then we described the changes that took place in geography during the 1950s and 1960s, concentrating on the rise of quantitative methods and spatial science, which together submerged much, though by no means all, of earlier geographic principles. In the course of the 1970s and 1980s the intellectual space created by the developments of the previous two decades, along with the increasingly evident shortcomings of quantification and spatial science, both allowed scope for and encouraged the emergence of a whole diversity of different approaches to geographical thought and practice. These approaches – which are the essence of this book – were briefly examined in turn, and then we concluded our chapter by turning away from such grand theoretical trajectories to the personalities and personality of geography as a lived entity, a subject peopled by conscious individuals. In the chapters that follow, these themes – of intellectual advance and debate, and the role of geographers as social scientists in actively fashioning geography within the constraints set by the conditions of the day – are elaborated upon at length.

Notes

1. Both broke with the practice established by Chorley and Haggett, separating physical from human geography because of a perceived drifting apart of the two branches, particularly through the forging ahead of human geography. For Peet and Thrift this meant no attempt to incorporate physical geography; the Gregory and Walford collection was accompanied by a parallel but separate volume,`*Horizons in Physical Geography*. (Clark, Gregory and Gurnell, 1989).

2. Throughout this volume one unfortunately enduring characteristic of the work we review – new and old alike – is a tendency to sexist language: the use of formulations that assume, quite erroneously, that all geographers are males, or that only males are of significance. Whilst we strongly oppose such insensitive and value-laden language, we have chosen not to remark upon it at every instance (such a task would have rendered the word *sic* necessary on virtually every page) but to highlight it in the form of a footnote here. In this sense perhaps it is unfortunate for Halford Mackinder that his should be among the first of many such quotations.

3. The debate that followed this paper, and is published alongside it, gives some further indication of the key issues of the day. Halford Mackinder (Herbertson, 1905, p. 309), for instance, commented that

> While I agree with Mr Herbertson in most of his ideas, I hope that the expression 'systematic geography' will not take root. This is an important point, and not merely

one of the words. Mr Herbertson's real aim, as it appears to me, is to obtain a method for 'Regional Geography'. . . . It seems to me a pity to confuse the public with a new term such as 'systematic geography' when you mean no more than can be conveyed by the term which some of us have been labouring to establish in current use.

The chairman of the meeting summed up proceedings (*ibid.*) as follows: 'To judge from my own experience, I should say that the varieties of nature are so infinite that I hardly conceive it possible to co-ordinate different parts of the world and to bring them together under one head as a type of regional geography' (p. 312).

4. Others were less concerned with regional *identification* than with the difficulties of *description*. W. M. Davis, for instance, whose descriptive device of structure, process and stage was decisive in the early 'exploratory description' of landforms, also commented as follows in an early paper (1915, p. 99) that presaged many subsequent debates, not least to do with ways of writing geography:

There can be little question that the least satisfactory feature of regional description lies in the necessity of presenting in separate, successive paragraphs or pages the many kinds of things that occur together in natural but unsystematic groupings. . . . Although all the geographical elements of a region occupy it continuously, they must be described in some reasonable sequence, for they cannot all be stated at once.

5. Sauer (1924, p. 18) once described his views on determinism as follows:

[My] position involves a conscious reaction from many of the older traditions of modern geography as fathered by Ratzel, which are derivatives from determinist philosophy. The argument from physical cause to human consequence, facile, alluring and plausible, rather curiously has had more vogue among British and American geographers than has been the case on the [European] continent. The tendency of this school of environmentalism is to seek out the so-called geographic influences with the purpose of determining the potency of the physical environment. It is difficult to do scientifically sound work under this plan.

6. Ackerman (1945, p. 141) concluded his remarks with the following programmatic statement:

Many of our graduates know a little about a lot. They and their studies easily give a justified impression of superficiality to outsiders, a confused illustration of our ultimate objectives, and an example which encourages all sorts of incompetents to try their hand at geographic interpretation. Human geography (and ultimately regional geography) will never be accepted as a mature scholarly discipline until a more thorough systematic literature begins to take shape in it.

Note also that Ackerman taught in the Department of Geology and Geography at Harvard University, the focus of a bitter (and ultimately unsuccessful) struggle for the establishment of a separate Department of Geography (considered later in the chapter).

7. Such concerns troubled, among others, Carl Sauer. In 1924 he wrote (p. 21) as follows:

It would be an interesting and disturbing experiment for several geographers to study the same area independently. Aside from the inevitable difference in quality of work, there would likely be serious discrepancies because of the lack of agreement as to the things chosen for observation, because of divergence of manner of observation, and finally because of differences of interpretation. Can we speak of scientific results until we have some measure of common discipline by which our work is unified? Have we not given excessive freedom to subjective impression, and in so far have been anti-scientific?

8. Hartshorne (1955) responded to this attack (which was published after Schaefer's death) with a sweeping condemnation, alleging that at least one section of Schaefer's article consisted 'in large part of false statements concerning the work of other students' (p. 205). He concluded (p. 243): 'In total, almost every paragraph, indeed the great majority of individual sentences in the critique, represents falsification whether by commission or omission'. Yet Hartshorne also recognised the essential validity of Schaefer's challenge against concepts that had previously been widely accepted. In a subsequent work (Hartshorne, 1959, p. 164) he largely agreed with much of what Schaefer had implied:

> Since geography requires both generic studies and studies of individual cases – it is in part nomothetic, in part idiographic – there seems little point in attempting to measure the relative amount of the two types of studies. . . . Each student may place his own emphasis on that type of study which he himself is most interested to pursue.

In a footnote to the same work (p. 165) he also maintained that a similar argument had been put forward in *The Nature of Geography* in 1939; in other words, he felt it was wrong to criticise even this earlier volume as wholly idiographic.

2

CHANGING TIMES AND THE DEVELOPMENT OF MARXIST APPROACHES TO HUMAN GEOGRAPHY SINCE THE LATE 1960s: STILL RELEVANT AND RADICAL AFTER ALL THESE YEARS?

Introduction

This chapter is concerned with the development of Marxist approaches to human geography in the period since the late 1960s. As already indicated in Chapter 1, the adoption and reformulation of Marxist ideas in human geography was only one of the directions that span out of the maelstrom of reactions against the constraints of spatial science. The initial basis of Marxist approaches to human geography lay in their oppositional critique of the limitations of spatial science and, in particular, its disregard for the economic and political constraints imposed upon spatial patterns by the way in which society worked, and its tendency to restrict analysis to how things actually *seemed to be* rather than to consider how they *might* be under different social conditions. This radical tradition, then, grew out of a dissatisfaction with existing analyses of – and roles for geography as – spatial science, and was involved with a search for alternative perspectives that were more socially responsible, or at the very least more aware of the social responsibilities attached to the pursuit of geographical research. In time, and with theoretical refinement, one element of this new critique became engaged with the social ideals and reforming bases of Marxist thought and practice. Our interpretation here suggests that this was a highly significant development; not just for its connection of geography with one of the main intellectual traditions of social science; nor even just for its reassessment of the relationship between geography as a discipline and the broader world; but also for its energising effect in laying the ground for many subsequent debates.

It is obviously impossible to do full justice in the space available here to the complex intellectual traditions of Marxism (see Figure 2.1) but a sketch of some of the main tenets of Marxist thought and their adoption in geography is nonetheless important. In considering urban spatial appearance, for instance, early radical geographic analyses (and one of the points we make is that these were not *necessarily* synonymous with or reducible to Marxist theory) saw evident patterns such as inequalities between different groups of people living in very different circumstances, in terms, for instance, of housing conditions

Figure 2.1(a)   The origins and different traditions of Marxism

Marxism originated in the nineteenth-century life and writings of Karl Marx and Friedrich Engels, in classic texts such as Marx's *Capital* and *The Eighteenth Brumaire of Louis Bonaparte* and Marx and Engels's *The German Ideology*. In recent years numerous guides and readers to the work of these authors have appeared, enabling and informing a whole generation of 'Marxologists' (see, for instance, Fine, 1975; Fine and Harris, 1979; Bottomore, 1973, 1981). There is, however, a highly complex intellectual genealogy not only to different stages of Marx's own writings but also to subsequent re-theorisations and practical applications. It is impossible to do full justice to this here, but it is none-theless important to acknowledge the diverse bases of Marxism – the different kinds of Marxisms – that have evolved. In part this spans the thoughts of Lenin, Stalin and other Soviet leaders, as well as the alter-native versions of Trotsky grounded in permanent revolution (see Man-del, 1979). It encompasses too a wealth of Western traditions, including the 'Frankfurt School' of critical theorists (see Figure 2.1(b)), the 'struc-turalist' versions of, for instance, Althusser (see Althusser, 1969; Alt-husser and Balibar, 1970) and the more 'humanist' and 'historicist' accounts of such as Thompson (see 1963). In recent years there has been a lively and varied debate within the Marxist tradition, much of which is mirrored in the geographical concerns considered in this chapter.

and quality of local environment. Studying the spatial expression of these inequalities led radical geographers to examine Marxist accounts of their social roots, especially in broader processes inherent to the capitalist system, which divided society up into classes that were highly unequal in terms of wealth, property and power. The most fundamental division in society – one at the heart of such inequality – lay between those who owned the means of production (the machinery and buildings of a factory, for instance) and those who had nothing more to sell than their labour power, the ability to perform paid work. This separation of society into two classes – capital and labour – rested upon an exploitative relationship in which capital gained more from the process of production than labour, via its appropriation of the surplus: its ability to pay labour a wage for subsistence and yet still retain the excess return of the value added in the course of production. This profit had to be re-invested in further, more exploitative ventures if the individual capitalist was not to be overtaken by competitors. In such an interpretation of history, the expansion-ary drive of capitalism would run eventually against its own limits, set by the contradictory nature of the class relation. For whilst capital needed labour as producers, it also needed to do away with it as far as possible and replace it by machinery (which was more efficient), thereby severing the link between pro-duction and consumption and leading to an economic crisis. In time and under such conditions, capitalist societies would therefore be transformed to a social-ist or communist state, in which the property relation was replaced with shared ownership of the means of production.

Figure 2.1(b)   The Frankfurt School and critical theory

> Critical theory is commonly regarded as a product of the Institute for Social Research, the first Marxist-orientated research institute in Germany, established in 1923. Its writings and arguments are often referred to as the Frankfurt School, which is unfortunate since much of the work was undertaken in exile in the USA, and in any case members of the school rarely shared any identifiable collective stance. The hallmark of critical theory is the way in which it identified the connections between social theory and social process in an attempt to transform the latter. It has been intricately connected with the modernist advance; for instance in the 1970s and 1980s one of its key exponents, Jurgen Habermas, sought to defend the project of modernity against the challenge of postmodernism (described in Chapter 6). Critical theory has had a complex and intricate two-way relationship with western Marxism, neither the sum of it nor reducible to it (see Kellner, 1989).

In these ways, then, radical or Marxist analyses of spatial patterns moved from a concern with inequality to investigate both its social *causes* (in the case of the city, for instance, through the workings of the land and property markets) and possible routes for the creation of alternative social forms that would render it impossible, let alone unnecessary. Such an interpretation of Marxist principles is of course highly over-simplified (if not that different from many of the thought-stages through which early Marxist geographers moved) and it is vital to emphasise at the outset the complex substance of and variety to the Marxist tradition. Clearly, the class relation was manifested in a variety of many different, often internally contradictory ways; clearly, societies did not develop from one mode of production (or basis of organisation) such as feudal to capitalist to communist in some kind of logical, inevitable progression; and clearly the role of the state in the organisation of society had to be considered. But these and other deficiencies notwithstanding (and see Figure 2.2 for an account of the main terminological issues involved in Marxist theory, as well as Figure 2.1), the engagement with Marxist ideas was vitally significant to the path of human geography in the 1970s and 1980s (and in a very real sense too the geographic heritage began to contribute also to Marxist theorisations in this period).

The importance of Marxism to geography was perhaps most clearly evident in the substance of the Marxist method – in particular, its groundings in the precepts of historical materialism, wherein human beings and social life were considered with regard to their broader relationship to prevailing social conditions of the day. For in materially relating analysis to possible alternatives, Marxist geography initiated (and was largely initiated in) a far-reaching critique of the practical consequences of geographic research (and the conditions under which it was undertaken) for perhaps the first time on any major scale in the history of the discipline. It is in this link between theory and practice, also, that much Marxist work is often evaluated (see, for example, the accounts of Third World revolutionary struggles in Wiles, 1982; White, Murray and White, 1983). One of the arguments that runs through this

Figure 2.2   Theoretical categories of the Marxist tradition

Marx's writings rested upon a *materialist* conception of history, through which he grounded the search for an understanding of general processes of social change in the bases of production. 'Production', so Marx argued, was essentially the process whereby human beings interact with and work on the objects of nature (such as land, water and minerals) to secure, in the first instance, an existence via the provision of food, shelter and other goods. And in the organisation of production, individuals entered into relations with each other, for instance, as capitalist and wage-earner, or feudal lord and serf, or slave and owner. These *relations of production* were the bases on which societies were constructed, and constituted the essential organising principle of different *modes of production* such as capitalism, feudalism or slavery. In this way an individual's place in society was severely constrained, being partly defined by existing social conditions in relation to the process of production. Such stability was an ephemeral phenomenon, though, for there was in every society (except perhaps the ultimate step of Marx's sequencing of modes of production, communism) an inbuilt tension between the *forces of production* (or its technological base) and the relations of production, wherein the former gradually and inevitably moved ahead of the latter, challenging the rationale of the whole system. Factory manufacture, for example, was difficult if not impossible under feudal conditions, leading to a breakdown of the feudal system and a move to a new mode of production, capitalism. Within capitalist societies the division of labour rested on a fundamental contradiction between capital and labour – considered in the text of this chapter – that carried the seeds of capitalism's transformation to a new, more egalitarian mode of production.

Such a theorisation (and, especially, our sketchy presentation of it here) is of course a highly schematised one, which tends to downplay the complexity of Marxist ideas. There is also a wealth of alternative analytical commentary from different branches of sociology, concerned with apparent deficiencies such as the technological determinism of Marx's account and its linear view of history; the assumed inevitability of certain changes; and the heavily structurally determined conception of society. Perhaps the most important debate, though, concerns the Marxist theorisation of class as the central organising principle within society, for many different versions of class theory have been advanced. Perhaps the most graphic defence of the significance of Marxist class theory in terms of its relationship to the *dynamics* of the overall system of production lies in this quotation from E. P. Thompson (1965, p. 357, emphasis in original):

> Sociologists who have stopped the time-machine and, with a good deal of conceptual huffing and puffing, have gone down to the engine room to look, tell us that nowhere at all have they been able to locate and classify a class. They can find only a multitude of people with different occupations, incomes and status-hierarchies and the rest. Of course, they are right, since class is not this or that part of the machine, but *the way the machine works* once it is set in motion – not this interest and that interest, but the friction of interests – the movement itself, the heat, the thundering noise.

chapter is that much recent Marxist geographic research has suffered by paying relatively little attention to the essentially political and practical aspects of the Marxist tradition.

This is partly to anticipate the story, however, for first we must trace out the complex genealogy of Marxist geography in the 1970s and 1980s, and its relation to both broader intellectual processes and to developments in the outside world. For if Marxism as historical materialism teaches that the links between analysis and practice are crucial, then it is imperative to situate our account of Marxist geography in the prevailing conditions of the day. At times, this will involve the imposition of what may appear to some as unnecessary simplicity upon the narrative, and it has to be accepted that there are and have been many different kinds of Marxist geography: just contrast the differing approaches of, say, Daniels (1989) with that of Peet (1975). It *is* possible, though, to discern broad movements in the history of Marxist geographical analysis – there are general undercurrents to research ideas and directions – and it is on this basis that the account below is premissed.

To simplify the presentation then, the chapter identifies four main phases in the development of Marxist approaches, largely but not precisely mirroring both the state of the global capitalist economy and dominant geographical concerns in the Western world. This correspondence is not coincidental for, as will be seen, a major influence on radical Marxist geography – perhaps not surprisingly given its emphasis on social change – has been the prevailing materialist conditions of the day. It begins in the later 1960s with social reform and political challenge, highlighting the emergence of a radical tradition from the anti-positivist movement, and charting the swing of one strand of this from liberal to Marxist critiques of society. By the early 1970s, this transition was more or less accomplished, and from then to the late 1970s, as global recession set in, a number of authors engaged upon the formulation and integration of a body of Marxist work within geography. This second period was characterised by a search for mature reflection. In the late 1970s such a quest increasingly drew Marxist geographers to a reconsideration of social theory. This engagement over questions such as the appropriate theorisation of the relations between agency and structure, and society and space, distinguished the third period. Then from the mid-1980s onwards, in the face of deepening global recession and avowedly neo-conservative government policies in many Western countries, and greater criticism from both within and outside the tradition, a degree of fragmentation occurred. The fragile consensus fought for and consolidated in the 1970s was shattered, and an uneasy uncertainty took its place. Marxist approaches in geography had established some recognition but precisely what they were, and indeed what they stood for, had in turn become problematic.

From liberal to Marxist geographic critiques of society

The 'mainstream' debate
The 1960s was a period of turbulent political change in many parts of the world, most especially in the USA. A heightened awareness of the unequal impacts of government domestic and foreign policies was at the root of a growing resurgence of socialist and radical politics, which had been dormant (indeed,

actively suppressed) during the long post-war boom. This engagement with such issues as the war in Vietnam, or civil rights, involved not just established political forms, but other less conventional ones, including many of America's university campuses. From these various poles, there developed a renewed and heightened criticism of the working of American society. Some difficult questions were asked: how, for instance, could the supposed 'land of the free' tolerate, indeed allow, inner-city ghettos and other, less obvious but equally significant, manifestations of the highly unequal distribution of income, with all the constraints that this entailed? Why was the USA still engaged in an evidently hopeless military campaign in such a distant theatre of war (namely, South East Asia) when social and economic problems abounded much closer to home? These and other issues had a pressing and urgent relevance as the decade drew to a close, not least to some within a discipline of geography that had been newly purged of previous consensus by the quantitative and spatial scientific revolutions, but whose new techniques and methods seemed to be capable of having little if any impact on either problem or solution.

The new radical current in American geography, then, had its origins in the material and social conditions of the time. We argue here, however, that in these formative years very little so-called 'radical' geography was indeed that radically different from the conventional and scientific geographies to which it professed such indignation. Whilst early progress was evident from some of the pages of the journal *Antipode*, founded at Clark University in 1969, it was difficult to distinguish more than a vaguely concerned edge to many interpretations of the crisis in geography's relationship to a broader world. The shift from liberal to more critical Marxist interpretations of the causes of the problem (as opposed to its manifest effects) took slightly longer to develop. This was partly due to the relative neglect of Marxist principles in geographic thought and practice – a situation that was in itself a signal comment upon the social roles played by a great deal of earlier geographic research, and its unquestioning engagement with national government objectives (and the reasons for and circumstances behind this limited encounter would certainly repay closer attention). Given this limited tradition of Marxist geographic research, in other words, it is perhaps not surprising that it took geographers in this era some time to wake up to (and of course accept) the full implications of their critique.

Richard Peet, one of the first and most prominent converts to a new radical geography, once identified two types of concern in these early days. There was, he wrote (1977, p. 244), 'among academically-oriented geographers an effort to change the focus of the discipline, from what were seen as eclectic irrelevancies, to the study of urgent social problems; among action oriented geographers, [a] search for organisational models for promoting social change'. For the former, he argued, 'relevancy' became the battle-cry; but in common with conventional and now-established scientific geography, investigation focused only on the surface aspect of the problem, not on the deeper underlying causes. As he put it, relevancy meant 'changing the topical focus of the discipline yet retaining the existing research methodology' (*ibid.* p. 245). One of the clearest statements of this liberal strand to the emergent radical geography was in an early essay by Morrill (1969/1970) on 'Geography and the transformation of society'. The problem, as he saw it (p. 20), was this: geography 'has been truly conservative, preferring to refine our understanding of spatial patterns of society as they are, than to question the "rightness" of these patterns, or the responsibility of society for them'. His solution specifically *excluded* the possibility of dramatic social change as the route to progress for, as he put it, 'the dreams of revolution are naive' (p. 21). Instead, he argued (in figures of speech that revealed just as much as the message itself), that the power of rational argument should be brought to bear on the populace in general and politicians in particular. His proposals for action saw a need for 'a convergence between capitalism and socialism' (p. 23), and his 'declaration of conscience' (p. 22) resonated with the ring of what might be called liberal professionalism (*ibid.*):

> we dedicate ourselves to a program of study, education and action, designed to influence more directly the patterns of our society, to create mechanisms for critical evaluation of public and private proposals, and to make our own positive proposals for improvement. As specialists in environment, location and area studies, we can no longer sit idly by while our world becomes less and less liveable.

In this way the role of the radical geographer was couched in broadly similar terms to that of the geographer as spatial scientist: still a technical expert, but now a more socially aware one.

The second strand identified by Peet involved a search for the means of social change. It was perhaps most clearly evident in the work of a one-time convert to 'theoretical geography', William Bunge. In 1967 he was refused academic tenure at Wayne State University and he began more actively to promote his alternative conception of the tasks of geography. Bunge saw a vital need to return to the one-time historical specialisation of the discipline, *exploration*: in this instance not of remote jungles or colonial interiors, but of the new unknown spaces, the inner cities of America. His 'Society for Human Exploration' was founded in 1968 and embarked on an expedition to Detroit. This developed courses in association with Wayne State University, specifically aimed at inner-city black residents; prepared reports for and with local community organisations on topics such as school decentralisation; and generally helped to provide an alternative source of information for low-income groups (see Horvarth, 1971). Bunge (1969, p. 35) set out his own ideas as follows:

> The purpose of the Expedition is to help the human species most directly. It is not a 'nice' geography or a *status quo* geography. It is a geography that tends to shock because it includes the full range of human experience on the earth's surface; not just the recreation land but the blighted land; not just the affluent, but the poor; not just the beautiful but the ugly. It is also a democratic as opposed to an elitist Expedition. Local people are to be incorporated as students and as professors. They are not to be further exploited.

In marked contrast to conventional planning, Bunge felt that the objective was to discover 'what the people in the region need geographically by becoming a person of that region' (p. 37). He recognised too that this was a far cry from the traditional career route in geography that, in his words, had 'nothing whatsoever to do with being oriented towards productive geography and everything toward "playing the game" of personal career' (p. 36).

Bunge's views on the geographical profession far from endeared him to many erstwhile colleagues (for a frank statement, see Figure 2.3). Yet beyond this it was difficult to identify much that was truly radical even about the

Figure 2.3   Bunge's views on the geographical profession (from Bunge, 1969, pp. 36–7)

At that point in their training where the student is supposed to do a significant piece of independent research, at last after starting in kindergarten as a total absorber of instruction, what does the typical geography graduate student do? He continues in his past pattern of trying to please his teachers. He cases the joint 'realistically' and rationalises his sellout with the slogan 'after I get my union card'. Having conditioned himself into seeing his research as the symbol of his lack of integrity, to say nothing of his manhood, that is, having sold his thesis for his degree, he simply continues this pattern the rest of his life. He publishes to keep from perishing. He sees tenure as the next 'union card'. And eventually he sees retirement as the goal of his existence. Along the way, he seeks out and finds a society of similar time servers, who rather than discussing what is wrong with themselves, the nature of geographers, they lash out endlessly, during marathon coffee hours, about the dismal nature of geography. Every academic geographer reading this feels the sting of the truth of these words. Armchair geographers of the world arise, you have nothing to lose but your middle-aged flab. It is not too late.

expeditionary movement. In the spirit of the day it was a bold and unusual departure from conventional norms: its principles professed commitment to change but its analytical commentary never developed in a way that moved beyond immediate objectives and into longer-term practical goals. Some of Bunge's comments on the nature of expeditionary activity clearly reflected his earlier interest in scientific geography: 'We must locate and map at least, and then spatially classify (regionalise) and then spatially predict the human condition' (*ibid.* p. 33). As Peet (1977) reflected, the relationship between Bunge's expeditions and 'a deeper, more all-embracing revolutionary movement was always tenuous at best, while at worst [they] might be considered a liberal diversion of political effort' (p. 248).

It was, therefore, partly because rather than in spite of the fact that the initial phrasing of questions to do with relevance and social change posed no really fundamental challenges about the working of society (and, indeed, Bunge apart, of the geographic profession, individually or collectively) that they came rapidly to occupy a prominent position in debate within the discipline. The intellectual ferment (which was heavily concentrated initially in American geography) was more widely conveyed to a British audience by two reports on the 1971 conference of the Association of American Geographers (AAG) (see Prince, 1971; D. M. Smith, 1971). These carried a similar message: that 'a new wind of change [was] beginning to blow, in the form of an emerging "radical" geography and an embryonic "revolution of social responsibility"' (D. M. Smith, 1971, p. 153). For Prince, this meeting of the AAG (held in Boston) reflected the climate in more ways than one. The atmosphere was chilly and the financial outlook for the geographical profession bleak. In consequence many geographers had begun to re-evaluate their contribution to society in other than strictly commercial terms, and to question the impact of the patronage of grant-awarding bodies in determining the research agenda.

For some, of course, there was no wish to join such a debate, and there were ample 'places of refuge' in sessions at the conference, which examined subjects either 'remote from the issues of present-day concern' or 'treated with a cold lack of moral sensibility and human compassion' in a 'game played by professionals, by hired mercenaries – but nevertheless, just a game' (Prince, 1971, pp. 152–3). For many, though, the signs of challenge were hard to ignore. The annual conference passed a motion calling for US withdrawal from South East Asia and a new organisation, the 'Socially and Ecologically Responsible Geographer', was founded by a one-time president of the AAG, Wilbur Zelinsky. To David Smith, the radical movement was 'calling for a greater professional involvement with matters of contemporary social concern' (1971, p. 154); part of a 'more general feeling that human progress needs to be accurately judged by social as well as economic criteria' (*ibid.* p. 156) – an early precursor of his own development towards a 'welfare geography' (see, for instance, D. M. Smith, 1977).

Even such limited statements of reformist intent met rapidly with a fierce response from some of the orthodox. Berry (1972a) replied to overtures of progressive reform with an attack on the 'new' agenda. He distinguished (p. 77) first between two different groups of radicals, neither of which was capable (in his view) of achieving social change:

> the majority of the new revolutionaries, it seems, are essentially 'white liberals', quick to lament the supposed ills of society and to wear their bleeding hearts like emblems or old school ties . . . . A smaller group of hard-line Marxists keeps bubbling the pottage of liberal laments. In neither group is there any profound commitment to producing constructive change by democratic means.

What was required of geography, he argued (p. 78), was *more* co-operation with the planners (along the lines of, for instance, Berry, 1970), *not* a questioning of their objectives: 'an effective policy-relevant geography involves neither the blubbering of the bleeding hearts nor the machinations of the Marxists. It involves working with – and on – the sources of power and becoming part of society's decision-making apparatus'. Yet as Blowers (1972, p. 291) identified in response, Berry's proposal created more problems than it solved, and effectively reinforced status quo attitudes: 'The issue is not how we can co-operate with policy-makers, but whether and in what sense we should do so. It is a question of values. Brian Berry's reflect those of an existing establishment'. From a radical viewpoint, Blowers argued (*ibid.*) that 'Research cannot be value-free for values are involved in the selection of research problems and in the evaluation of results. What matters is that the values upon which the research is based are made more explicit'. Through this simple expedient, Blowers suggested, the discipline would be better equipped 'to influence social reform' (p. 292).

But at just the same time that the call for social reform was becoming accepted by a sizeable minority within the discipline, and the focus of a challenge from the old-guard, a transition was taking place within the radical current, carrying some at least away from liberal perspectives and into more explicitly Marxist interpretations of society. A central turning point in this process was a paper by David Harvey, one of the foremost codifiers of the quantitative school, who had since turned his back on such concerns. His earlier achievements within the scientific revolution (in particular, *Explanation in Human Geography*, 1969b) meant that he could not easily be ignored; his personal contribution to the development of Marxist perspectives in geography was (and remains) deeply significant.

In 'Revolutionary and counter-revolutionary theory in geography' he explicitly addressed the question of ghetto formation from a radical stance (see Harvey, 1972). The first sentence posed a seemingly simple question: 'How and why should we bring about a revolution in geographic thought?' But in answering it Harvey opened several new and potentially challenging directions. For he located social science directly in the materialist world: not independent from but intimately related to existing social (and, in particular, class) relationships. Neither society, nor social science (and still less geography), he argued, were coping very well; as evidenced by poverty, war and disease on the one hand, inadequate solutions on the other: 'It is the emerging objective social conditions and our patent inability to cope with them which essentially explain the necessity for a revolution in geographic thought' (*ibid.* p. 6). This new revolutionary theory would need to take full account of 'the social bases and implications of control and manipulation' (p. 4). But Harvey also identified another possible route for social science, that of *counter-revolutionary theory*. This, he argued (*ibid.*), 'is one which is deliberately proposed to deal with a proposed revolutionary theory in such a manner that the threatened social changes which general acceptance of the revolutionary theory would generate are, either by cooptation or subversion, prevented from being realized'. He went on (*ibid.* p. 10) to illustrate precisely what this meant by describing what true revolutionary theory did *not* encompass:

> It does not entail yet another empirical investigation of the social conditions in the ghettos. We have enough information already and it is a waste of energy and resources to spend our time on such work. In fact, mapping even more evidence of man's patent inhumanity to man is counter-revolutionary in the sense that it allows the bleeding-heart liberal to pretend he is contributing to a solution when he in fact is not.

In other words (p. 11),

> the emergence of a true revolution in geographic thought is bound to be tempered by a commitment to revolutionary practice. Here there will be many hard personal decisions to make. Decisions that require 'real' as opposed to mere 'liberal' commitment, for it is indeed very comfortable to be a mere liberal.

In making explicit this connection between theory and practice, and in signalling the counter-productive work of liberal reform from a revolutionary stance, Harvey effectively shifted the terrain of debate within radical geography on the one hand, and in the response to radical geography within the discipline on the other. Some positively welcomed his call for a 'Marxist geography' (see Folke, 1972); others, largely from outside the radical current, were much more critical. Berry (1972b) took a diametrically opposed view of the ghetto, Harvey's starting point for analysis, arguing (p. 33) from a pluralist perspective on society that:

Figure 2.4   David Harvey's Social Justice and the City *(1973)*

David Harvey's *Social Justice and the City* (1973) was a highly significant statement to the newly emergent Marxist tradition within geography. It was set in two main parts, hinged around a version of
Harvey's essay on 'Revolutionary and counter-revolutionary theory in
geography . . .' (1972). In Part I Harvey focused upon liberal formulations of the urban problem, examining issues of distribution and justice
within the urban environment. Then in Part II he expanded upon more
explicitly Marxist principles, first with the statement on revolutionary
theory, then introducing and synthesising Marxist concepts to do with
use-value, exchange-value and modes of production to the urban debate. Essentially the book represented a report of work-in-progress
rather than a definitive statement, for the essays it contained 'were
written at various points along an evolutionary path and therefore represent[ed] the history of an evolving viewpoint' (1973, p. 10). It was in
Part II that 'some fundamental lines of thought and ways of thinking
[were] opened up' (p. 18). In this part of the book, four concerns related
to social process and spatial form were subjected to Marxist analysis
and explanation – the nature of theory, space, social justice and urbanism. The last three of these became the focus not just as objects in their
own right (as they had been in Part I) but as intimately bound up with
broader social processes. 'Urbanism', for instance, ceased to be a thing
in itself and became instead a vantage point on society. In this part too
Harvey rejected the separation of methodology from philosophy that, he
argued, had underpinned his *Explanation in Geography* (1969b). The
book's conclusions, in a brief Part III, dealt with the significance of
Marxist method to the analysis of urban problems. Ultimately though,
Social Justice and the City was important not so much for what it said
(although much of this was novel to geography) but for the avenues of
future enquiry it opened up, and as a statement of the potential power of
Marxist explanation.

bubbling Black ethnocentrism, with an alternative twentieth-century core of theories
of social change, provides an alternative value system that provides a fundamental
basis for optimism. In this view the ghetto is not a problem; its accelerated growth
and zones of abandonment are opportunities for acquiring political self-control and
for using the power base to effect institutional change.

Harvey later built on this earlier paper to make explicit his transition from
liberal to socialist formulations in the influential *Social Justice and the City*
(1973) (see Figure 2.4). This drew another fiercely strong response from Berry,
indicative of growing personal tensions raised by the debates (see Figure 2.5).
Other authors though were beginning to develop work in the Marxist vein; for
instance, Anderson (1973) argued for a close examination of the role of ideology and a regrounding of geography in the workings of capitalism from a
Marxist perspective. Lacoste (1973) was also centrally concerned with the role
of geography in society in his account of 'geographical warfare' via an analysis
of bombing patterns in Vietnam. He found 'proof of the deliberate and

Figure 2.5 *'Berry* v. *Harvey': selections from a debate over Harvey's* Social Justice and the City *(from* Antipode, *1974, pp. 142–9)*

> Berry, in reviewing Harvey's book, described first what he regarded as its prime contents. He delivered judgement:
>
>> Harvey is stimulating, provocative and even on occasion profound. And – as far as I can tell – he is consistent within his chosen assumptions, although he may sit a long time waiting for the revolution . . .. And yet . . . the functional basis of the system of industrial capitalism is changing . . . [through] the subordination of the economic function to the political order, thus providing the formerly-missing economic controls . . . while Harvey is waiting for the transformations that will produce more than just forms of industrial capitalism, the actual transformations will have produced the post-industrial world.
>
> Harvey began his reply as follows:
>
>> Brian Berry writes a lot of reviews. And like mass production of any kind, such reviewing requires a formula. Such formula reviewing rarely leads to cogent or trenchant criticism and I know several editors who swear they will never send another book to Berry again because of it.
>
> He went on:
>
>> Berry appears to be saying that society has been so transformed by these 'new' control systems that the lineaments of a new society are already present – all of which, presumably, means that Marx is quite passée. But Berry's remarks are political rhetoric, not cogent criticism . . .. The only conclusion I can reach is that Professor Berry, whether he knows it or not, is simply defending the 'ruling ideas of the ruling class'.
>
> Berry continued the exchange:
>
>> Marxists become very testy when a critic refuses to discuss the internal details of their argument and argues instead that the premises on which the argument is based are false. Harvey appears exceedingly testy. He bewails the fact that I did not respond to some of his specifics. I did not do so quite deliberately, because I believe . . . his specifics are products of a particular mind-set that produces axioms of appeal only to the New Left elitists of the 1960s, the self-appointed spokesmen of the revolutionary counter-culture.
>
> Harvey:
>
>> It is fairly evident that Berry is neither interested nor capable of discussing these issues rationally.

systematic nature' (p. 8) of attacks on dikes in the Red River, geographically clustered so as to cause maximum flooding and damage to the civilian population. As he put it (p. 1–2),

> It is important that we gain (or regain) an awareness of the fact that the map, perhaps the central referent of geography, is, and has been, fundamentally an instrument of power . . . . Many university geographers today, who regard their discipline as a science – detached scholarship dedicated to knowledge for the sake of knowledge – would no doubt be shocked to learn the military and political nature of geographical thought.

By now the debate over relevance, an 'applied' geography and the questions raised implicitly or explicitly by the radical current, had come to prominence on both sides of the Atlantic. The 1974 annual conference of the Institute of British Geographers was the first ever to have a theme: 'Geography and public policy'. The case for greater involvement by geographers in public affairs was put in the President's address (Coppock, 1974). He argued that the discipline could ill afford not to become so involved, since that would lead to a decline in its importance with respect to other subjects. He recognised but did not accept the implications of the radical critique (*ibid*. p. 8): 'Fears have been expressed that applied research which is policy-orientated will be harmful to the profession, distorting academic judgements and prostituting the profession. My own view is that such research is both necessary and beneficial'.

Harvey (1974b) responded directly to these assertions. He set out two related questions: what kind of geography, and what kind of public policy? In answering these he focused on the reasons for geography's engagement with public policy. He discounted personal ambition and concern for the reputation of the discipline as too simple, and concentrated instead on social necessity and moral obligation. Social necessity he identified in the rise of the corporatist state during the 1960s, whose slogan was 'the national interest', for which a 'technically proficient bureaucracy' had to be produced. In responding to its needs and priorities (such as the preservation of economic growth, management of cyclical economic crises and containment of discontent) geographers had become more subservient to the state: 'In the process we have learned to be good citizens, to prostrate ourselves and prostitute our discipline before "natural priorities" and "the national interest"' (*ibid*. p. 21).

The only solace to be gained for Harvey from this state of affairs was in a growing sense of moral obligation. Whilst this might appear as counter to the ambitions of a corporatist state, such tensions *could* be accommodated by separating 'fact' from 'value', as achieved in the scientific revolution. For Harvey this was not enough. Issuing a clarion call for further reflection on and criticism of the basis of power in society, he concluded that the moral responsibility of the geographer should become social necessity of a more critical kind. Through such analysis, the geographer should 'attempt to subvert the ethos of the corporate state from within' (p. 24).

In many senses this debate over relevance and public policy, and publication of *Social Justice and the City*, represented landmarks in the contribution of Marxist analysis. A clearly defined and cogent Marxist geographical critique of society had begun to take shape. Although the overall numerical significance should not be over-emphasised, a discernible oppositional movement was beginning to emerge. Yet whilst the 'mainstream' of debate had run a clearly defined course, albeit with twists and turns, it is important not to lose sight of the other alternative directions that had spun out of the whirlpool of change in the late 1960s and early 1970s; nor to forget that for many in the discipline it was very much a case of business (literally) as usual.

Secondary currents and whirlpools
William Bunge, for one, grew increasingly marginal to the academic world, though no less committed to his cause. The findings of his expedition to Fitzgerald, a neighbourhood in Detroit, were reported in detailed form in Bunge (1971). In the Foreword to this carefully illustrated descriptive account

he repeated his earlier concern to establish a new kind of relationship between researcher and researched, one that broke down oppositional barriers: 'A regional study must be done by a geographer who calls the region home. It is impossible to understand a neighborhood without becoming a neighbor'. He went on to consider the broader significance of his work in Detroit, in a comment that both prefigured many later debates and reflected his earlier interests (*ibid.*):

> [A] major technical difficulty is the generalizability of the story. What can we learn about America from studying a core region of only one mile? Instead of following the geographical tradition of learning little things from a little region, this study attempts to learn big things from a little region . . . [this] reflects the author's deep feeling that locations are relative and not unique, that inferences can be drawn from samples.

Departure from the USA brought to a close the Detroit Geographical Expedition but, newly arrived in Canada, he founded the Toronto Geographical Expedition (see Stephenson, 1974). His writings came to focus on the geography of human survival in the face of the threat, particularly to children, from 'machinekind' (see Bunge, 1973). His goal, increasingly, was to ensure that 'mankind comes to some peace, some rest, some harmony, and ultimate unity with nature' (*ibid.* p. 290).

Others dissatisfied with the strictures of conventional and/or scientific geography developed ideas from similar bases of critique as the radicals, but expanded them in markedly different (if no less productive) directions. Among these, two of the most prominent were Hägerstrand and Olsson. Hägerstrand (1970) moved from a concern with people in regional science to the early formulation of time geography, of paths constrained in time–space and of 'daily prisms' (later to become one of the points of contact between human geography and structuration theory – see Chapter 4). His conclusion (p. 21), though, could equally have come from the liberal critique of the day: 'The technological forecasts which edify us these days and often seem so promising, at least superficially, cry out for instruments which could help us to judge the impacts on social organisation and thereby the impact on the ordinary day of the ordinary person'. Olsson, via several contributions to *Antipode*, developed a critique of ideology and methodology in spatial planning, and considered the relation between spatial analysis and geographic structuralism (see Olsson, 1972, 1974a, 1974b). In the light of his own transition from one school to another, he began to question the language of reasoning. He argued that the spatial analyst was a 'prisoner' within a certain 'reasoning mode'. He went on (1974b, p. 53): 'The lesson is that our statements often reveal more about the language we are talking *in* than the things we are talking *about*'. But from *both* perspectives he was convinced that 'we cannot increase our understanding of human behavior without drastically changing the language in which we think and talk about it' (*ibid.* p. 54). Such concerns were, again, a clear forefunner of much of his later work (see, for instance, Olsson, 1980).

For a large number, of course, the rise of radical geography was more or less an irrelevancy. Many of the traditionalists and the one-time 'new' scientific geographers could at least agree on this (see, for instance, Wilson, 1972). For still others, the radical debate engendered concerns that were to spin off in other directions, perhaps tangential to the main Marxist current but nonetheless still relevant to its course. The reasons for such peripherality were diverse. Eliot Hurst (1973, 1980), for example, moved from a critique of 'establishment geography' to argue for a 'de-definition' of geography. This would produce 'A completely new and radical . . . focus . . . that is grounded in the praxis of Marxism, but which transcends all of the current "divisions" of epistemological space' (Eliot Hurst, 1985, p. 85). Such a concern had been prefigured by Harvey (1973, p. 147) in his claim that 'the organisation of knowledge (including the disciplinary divisions) has an inherently *status quo* or counter-revolutionary posture'. But, of course, transcending disciplinary boundaries posed enormous problems for those working within them; and whilst inter-disciplinary work blossomed, a sustained challenge to the boundaries between subjects, and their rationale, was never seriously mounted.

In a related fashion it took the early radicals relatively long to recognise the heritage of anarchist tradition from within geography, left by such as the early Russian geographer, Peter Kropotkin (see, for instance, Galois, 1976; Kropotkin (reprinted), 1978; Breitbart, 1978/1979, 1981) and anarchist politics never really gained a wide circulation. Similarly, one of the most vibrant research areas of the following decade, the feminist critique, was surprisingly under-represented in early radical geography. Writings on the role of women in society, for instance, were scarce (for an exception, see Hayford, 1974), and the eye of the present day reader often looks rather warily at the gendered language in many radical texts. This, too, was an area of limited development.

Despite this, though, there was no doubt that Marxist geography embarked on the remainder of the decade with a sense of purpose. In a number of related fields clear research lines were being laid out and new avenues investigated. These included the study of development and imperialism; the process of uneven regional development; and theories of the city and urbanisation. These more fruitful research paths were intricately bound up with the mainstream current of debate. The following section therefore continues the historical narrative, taking the story of Marxist approaches in geography on from the early 1970s through to the late 1970s, via an analysis of these research areas, where necessary tracing their heritage back to earlier days.

The search for mature reflections

Once again, the pages of *Antipode* provide a useful indication of the state of Marxism within geography during the period from the mid to late 1970s. For instance, a seminar held at Clark University in 1974 on 'Marxist geography' drew around fifty people – a revealing estimate of contemporary support. Onset of recession in the Western world, along with political change in the USA, suggested hard times were ahead for revolutionary struggles both within and outside academia. As Bunge (*Antipode*, 1975, p. 83) put it in a letter to the gathering,

> Some years ago I opposed rump meetings since I thought they would be tiny and we still had a lot of wiggle room inside academic geography. I think the situation has changed drastically under Adolph Nixon's ruling-class-directed hand. The campuses are now as far to the right as they were to the left a few years ago.

It would be wrong to give the impression though, that all was retrenchment in this period. Whilst there were undoubtedly internecine disputes (perhaps

Figure 2.6   The Union of Socialist Geographers (see Dear, 1975)

The Union of Socialist Geographers was founded in Toronto in May 1974. Its goals were:

1. organising and working for radical change in society; and
2. developing geographic theory to contribute to revolutionary struggle.

The most fundamental change it sought was the 'socialization of the means of production', as the route to a 'humane, non-alienating society'.

At the first annual meeting in Vancouver in 1975 it was proposed to organise a 'field-trip to working-class Vancouver'. This drew a strong response from Michael Dear who argued that a truly socialist geography was *not*:

1. 'academic tourism of the kind which conducts tours of working class districts (such tourists could justifiably expect to be met with a punch on the nose)';
2. 'academic posturing';
3. 'a personality cult';
4. 'paternalism or elitism';
5. 'a people's geography'; and
6. 'using class origins as credentials'.

But that it should be 'an open attempt to formulate a new critique of society and its institutions'.

---

inevitably so given the emerging status of the new focus within the discipline – see Figure 2.6), several important research areas were mapped both in the USA and, increasingly, in Britain.

Among the earliest concerns of Marxist geographers was analysis of the twin processes of development and imperialism in the Third World. Here, drawing on the work of non-geographers like Frank (1971) and Amin (1976, 1980), a strong research tradition was quickly established. For instance, Blaut (1970) criticised the construction of a Western-European centred 'ethnoscience' and its impact upon understanding of the Third World, calling it the 'intellectual underpinning of imperialism'. He subsequently wrote widely on imperialism and development (see 1973, 1975), locating the former in the geographical expansion of capitalism and continuing his attempt to 'de-Europeanise' theories of development. Such concerns were further developed by the Brazilian, Santos (1974, 1977), and by Slater (1977), whilst the early writing of Buchanan (1974), including his reflections on the 'dirty word' of development, slowly gained a wider audience.

Two particularly significant papers extended analysis of the ideology of development into broader questions. Hudson (1977) located the rise of geography from 1870 to 1918 firmly in the context of the geographical expansion of the colonial powers and their pressing need for maps and surveys to underpin and codify white rule. The context and status of geography in this period clearly reflected its relation to a dominant, aggressively expansionist, political conception. Blaikie (1978) questioned Western conceptions of the development process, through an analysis of prevailing ideas about spatial diffusion as a complement to 'modernisation'. In order to understand space, he maintained, 'a theory of social change is required' (p. 270). This social theory and its 'particular manifestations' were themselves ideological; therefore diffusion theory was at fault for its failure to consider the bases of social change. In these and other ways, Marxist geographic analysis of development processes signposted valuable areas for future research into the relationship between the geographic discipline and the material world. It also highlighted early tendencies towards the use of functionalist explanation wherein geography seemed to take on certain roles because it was socially functional for it to do so. This was a recurring theme, to which we return below.

A second expanding line of research concerned analysis of uneven regional development, primarily in the Western world. In an early paper, Doreen Massey (1973) initiated a critique of industrial location theory, questioning its orthodox separation from analysis of the economy. She concluded that 'most of existing industrial location theory is placed within an ideology which defines its object and mode of analysis in a way which makes effective analysis impossible' (p. 39). From this reference point several workers set out to develop an alternative conception of industrial location, which could integrate it into the functioning of society. In a later article, Massey (1978) broadened the debate to encompass whole patterns of uneven development. She began with the argument that production and geography were intrinsically inter-related to outline a conceptualisation of 'rounds of investment', in each of which a new form of spatial division of labour was evolved. Consequently, 'the social and economic structure of any given local area will be a complex result of that area's succession of roles within a series of wider national and international divisions of labour' (*ibid.* p. 116). This 'spatial division of labour' approach was developed in subsequent work (see Massey, 1979, 1983, 1984). Other researchers undertook, or initiated, wide-ranging investigations of the precise nature of change in particular regions (see, for instance, Carney, Hudson and Lewis, 1976, 1980).

Third, research increasingly concentrated upon urban environments. Here, Harvey continued to develop his Marxist critique of society, focusing for instance upon the structure of the housing market in Baltimore (Harvey and Chatterjee, 1974). In the UK several researchers developed similar concerns; for instance, Boddy (1976) examined the role of building societies in structuring urban space. Urban geographical research was partly informed by (and partly oppositional to) the work of French sociologist, Manuel Castells (see, for instance, 1977), and several formulations of the significance of urbanisation were developed (for instance, Harvey, 1978). But despite the increasing sophistication of later theoretical work, the urban context exhibited most clearly of all research avenues one particularly difficult problem that was increasingly confronting Marxist analysis. This was the issue of how to relate the observable structures of society at a macro-scale with the events and outcomes of day-to-day life at a much smaller scale.

Much early Marxist analysis fell into the twin pitfalls of structuralism and functionalism, arguing implicitly or explicitly that the structures of society dictated actions and events almost behind the backs of the individuals concerned, and that society operated in the way that it did because it was necessary for society's survival or functioning. Both positions were clearly bound up with the way in which early Marxist geographic research had been initiated in a critique of spatial science's disregard for the structural determinants of spatial patterns, on the one hand, and its attention to how things were rather than how they might be, on the other. Equally, such views – or early starting points – proved to require considerable re-evaluation in the light of experience and counter-critique. For instance, Peet (1975, p. 564), in an attempt to develop a Marxist-geographic theory of poverty, tried to synthesise two concepts: 'the Marxist principle that inequality and poverty are inevitably produced by capitalist societies, and the social-geographic idea that inequality may be passed from one generation to the next via the environment of opportunities and services into which each individual is implanted at birth'. In this synthesis he saw the role of specifically Marxist theory as 'dealing with the great forces which shape millions of lives' whilst 'geographic theory deals with the mechanisms which perpetuate inequality from the point of view of the individual' (p. 567). Thus Marxism was about the way society worked whilst geography was about the patterns society produced and their impact on the individual. There was little or no scope for the individual to have an impact in changing society – totally contrary to both common sense and to many Marxist interpretations of social change. Such problems were to be the motivation behind a shifting emphasis in Marxist geography as it sought to respond to the humanist critique that the Marxist geographic emphasis on structure and constraint was over-played at the expense of human agency and opportunity. In this broadening debate, Marxist geography looked more generally to social theory both for the right questions to ask, and for some at least of the answers.

Engagement with social theory and reconsideration of the role of structures

If the (belated) coming of age of Marxism within geography was in a sense marked by the publication of a collection of essays from *Antipode* (Peet, 1978), it was a difficult early adulthood. The critique of structural Marxism from both within and outside was a trenchant one. In this debate many issues were raised, and the relationship between geography and Marxist theory was re-evaluated. The general development of humanist concerns in geography is considered in Chapter 3. We focus in this section first upon the humanist critique of Marxist geography, then on some of the directions in which Marxist geographers responded to criticism both from without and from within the Marxist tradition during these years.

In a provocatively titled essay, Eyles (1981) argued why in his view geography 'cannot be Marxist'. His argument was that 'for an understanding of lived experience, it is necessary to transcend some of the orthodoxies of Marxism, particularly those of structuralist Marxism' (p. 1371). The problem with the latter was that it treated lived experience 'as a determinate outcome of structural processes, or more crudely, structure is seen as determining behaviour' (p. 1372). What was needed, he argued, was a transcendence of Marxism by a more humanist approach, encompassing the socially constructed world but still grounded in material existence.

Duncan and Ley (1982) also rejected structural Marxism. Their concern was three-fold: the status of individuals in regard to social structures; a tendency to functionalism; and the relation to empirical evidence. On the first count they reiterated Eyles' concerns, arguing that structural Marxism involved reification – giving mental abstractions (like class and the state) undue causal powers. On the second, they identified the tautology of functionalism, wherein a function exists because it is functional, and it is functional because it exists. This logical error, they claimed, characterised much Marxist geography. And on the third they discussed the problem of identifying 'classes' as one instance of a more general tension between structural theory and empirical investigation, and the inability of the former to accommodate lived experienced.

This was responded to by Chouinard and Fincher (1983), who argued that Duncan and Ley had ignored the development of a variety of different Marxist approaches since the mid-1970s. Not all contemporary Marxist geography, they argued, was structuralist in content or intent, and it was wrong to characterise it as such. In defence of Marxist theory, Chouinard and Fincher also identified what they saw as gaps in Duncan and Ley's logic, including ignoring the significance of class struggle as a route for agency and social change. They concluded (p. 144) by highlighting Duncan and Ley's failure to specify an alternative theoretical basis, accusing them of foreclosing constructive debate on the issues raised, adding in no uncertain terms: 'we would hope that future "critical analyses" of Marxist work in geography might offer more opportunities for a constructive exchange of ideas between social theorists'.

Chouinard and Fincher were certainly right to argue that there had been a lively debate within Marxist theory in the second half of the 1970s, focused in particular upon a reappraisal of the structuralist legacy of Althusser (see, for instance, Althusser and Balibar, 1970). There were two clear expressions of this: one in a reconsideration of the appropriate theorisation of the relation between society and space, and a second in analysis of the relationship between agency and structure. Together, these encapsulated growing moves by Marxist (and non-Marxist) geographers to engage with contemporary social theory.

If spatial science had been concerned with spatial structures and the (spatial) relations between them, then Marxist geography in particular developed from an interest in the inter-relations between social relations and spatial structures to re-evaluate some core geographic and Marxist ideas about the significance of space. Soja and Hadjimachalis (1979, p. 4) asked one of the fundamental questions in this debate: 'Must historical materialism be complemented by an equally robust geographical materialism?' And, in the course of answering in the affirmative, commented (p. 9) that

> Marxist spatial analysis [and note the use of this term] today is beset with several internal divisions which we feel can and need to be overcome: . . . [The most basic of these is] between those who see the spatial problematic as an important, but easily fetishized [that is to say, granted inordinate and undue status] contribution to historical materialism, and those who see it as the basis for a geographical and historical materialism dialectically intertwined.

Their concern was with the need to recognise and defend the significance of space to an historical-materialist approach against the charge that social relations were constitutive of – that is, determined – spatial patterns and processes (the shadow of geography's earlier struggle to legitimacy was an intriguing and partly ironic one here, especially given the Marxist tradition's disavowal of disciplinary boundaries). Theorists such as Harvey had criticised this view of the social *production* of space for denying the importance to society of spatial factors. Soja (1980) maintained also that it was important not to lose sight of the essentially dialectical character of the relationship between society and space. In his words, 'the general relations of production' should be conceived as 'simultaneously social and spatial' (p. 208), and as such formed the socio-spatial dialectic.

To some Marxist geographers, however, it was still necessary to maintain the integrity if not the independence of the spatial. Peet (1977, p. 254), for instance, argued that 'geographic variation gives a strong spatial weight to social process, so much so that we can speak of spatial processes'. This brought a sharp rebuttal from Neil Smith (1979, p. 376), that 'far from being different relations, the social and the spatial are different aspects of a single relation'. Peet (1981, p. 105) continued the debate by defining 'spatial dialectics' as part of 'the attempt by Marxist geographers to move through spatial description into an analysis of the social processes which originate spatial appearance'. Whilst recognising that 'social process and spatial form are thus components of a dialectical unity' (p. 106), he none the less insisted that 'we would miss an entire dimension of understanding if we did not also grant to spatial relations a relatively autonomous position, a history in part of its own, with a dialectic in part of its own' (p. 108). In reply, N. Smith (1981, pp. 112–13) argued again that

> In fact, there is nothing dialectical about 'spatial dialectics'. It begins from a dichotomy that it never does transcend: space on one side, social process on the other. The theory conceptualises space as separate in the beginning from social process in order to be able to relate the two dialectically, but this dialectic never comes to fruition. Far from achieving a dialectic, the theory treats space in practice as a relatively autonomous thing or field . . . the product is in practice little more than a banal reaffirmation of the idea that space and society interact. Crude spatial interactionism masquerades as spatial dialectics. It refines but does not remove the orthodox fetishism of space.

As even these limited exchanges reveal, then, the reassessment of the significance of space to Marxist geographic thought was a far from simple or consensual process.

A second component of the engagement with social theory concerned the question of the relationship between agents and structures (and here we touch briefly upon issues discussed at length in Chapter 4). In an early study Gregory (1978) provided a wide-ranging account of critical social theory (see Figure 2.1) that attempted to introduce German theorists such as Horkheimer and Habermas (see, especially, Habermas, 1976) to the debate in geography. The concepts of systems theory, for instance, were analysed for their collective interest in technical control (Gregory, 1980). He subsequently developed these ideas, drawing in part on the work of Anthony Giddens (1979, 1981) to outline the theory of structuration for a geographical audience (see Gregory, 1981).

Thrift (1983) sought also to integrate these ideas, identifying four concerns of the structurationist school that, he argued, were central to a non-functionalist

Marxism. It was avowedly anti-functionalist in intent. It recognised that social structures are characterised by their 'duality' – 'both constituted by human practices, and yet at the same time the very medium of this constitution' (p. 29). It incorporated a theory of action. And finally, it took account of the centrality of time and space to the construction of all social interaction and, consequently, in the constitution of social theory. In the process he tried to show (p. 38) that 'there is a place for a reconstructed regional geography, a regional geography that builds upon the strength of traditional regional geography, for example the feel for context, but that is bent towards theoretical and emancipatory aims'. This argument was mirrored by Gregory (1982c, p. 189), who argued that a central task was to 'underscore the importance of concepts of spatial structure to the human and spatial sciences as a whole'.

The effectiveness of geography's engagement with social theory was neatly encapsulated in a later work of Giddens (1984), who had himself (pp. 110–11) drawn on the work of time-geographers such as Hägerstrand:

> Over the past few years there has taken place a remarkable convergence between geography and the other social sciences, as a result of which geographers, drawing upon the various established traditions of social theory, have made contributions to social thought of some significance. Most such writings, I think it would be true to say, remain unknown to those working in the rest of the social sciences although they contain ideas of a very general application.

Sociology, he argued, could learn from geography about the significance of place in the reproduction of social practices. There were, after all, 'no logical or methodological differences between human geography and sociology' (p. 368).

So Marxists within geography, and geographers more generally, had begun to broaden their horizons to some effect, introducing refined formulations of agency/structure and society/space into their work, in the process posing fresh questions of Marxist theory. By this time a clearly defined body of Marxist geographic theory was being developed and a range of empirical research was underway. The most significant codification of theory came with the publication of David Harvey's *The Limits to Capital* in 1982 (see Figure 2.7), which was followed by a clarion call for further engagement with Marxist theory in geographical research (see Harvey, 1984). In this address to the annual conference of the Institute of British Geographers, Harvey (*ibid.* pp. 6–7) called for a continued commitment to theory, identifying also the challenge and the dilemma of contemporary Marxist theory and practice:

> Those who broke out from behind the safety of the positivist shield ruptured the political silence within geography and allowed conscience and consciousness freer play. But they spoke with many voices, generated a veritable cacophony of competing messages, and failed to define a common language to voice common concerns. . . . Between the safety of a positivist silence and the risk of nihilistic disintegration lies the passage to a revitalised geography . . . .. How to negotiate that passage is our dilemma of this time.

Navigation of this route was to prove a difficult task in the socio-economic environment of the mid to late 1980s, and its mediation in the realm of politics, as the following section goes on to show.

Figure 2.7   David Harvey's The Limits to Capital *(1982)*

David Harvey's *The Limits to Capital* (1982) represented a highly refined Marxist geographical statement of theory. In one sense it was the
culmination of questions Harvey had begun to ask in his earlier works.
As he put it in the introduction (p. xiii), 'After the completion of *Social
Justice and the City* (nearly a decade ago) I determined to improve
upon the tentative, and what I later saw to be erroneous, formulations
therein, and to write a definitive statement on the urban process under
capitalism from a Marxist perspective'. In this task though, mainly because of the inter-connectedness of Marxist theoretical categories ('It is
both a virtue and difficulty in Marx that everything relates to everything
else' – p. xiii), it proved necessary instead to write a general treatise on
Marxist theory, paying special attention to the dynamics of urbanisation.
Limits to Capital was the result. It contained three separate but related
'cuts' at a theory of capitalist crisis. The first rested upon the fundamental contradiction between capital and labour and the tendency towards a falling rate of profit. In the second, the role of the credit system
was introduced as a means of (temporarily) resolving contradictions in
the process of production, by stabilising capital for future rather than
present use. The financial system could only internalise capitalism's
tendency to crisis, though, leading ultimately to panic and inflation.

It was in the third 'cut' that Harvey introduced most specifically the
geographical aspect to capitalist crisis, focusing on the system's *spatial*
as well as *temporal* dynamics. For, he argued, the contradictions of
capitalism were susceptible to a 'spatial fix' (as well as a temporal one
through credit) in which geographical expansion into new regions temporarily ameliorated crisis tendencies, as part of the process of uneven
development. In this way capitalism was capable of 'switching' crises
from one region to another to smooth the overall process of accumulation, with inevitably devastating results for some places and regions.
However, as Harvey put it, the problem for capitalism was that the more
it developed as a system, the more it tended to succumb to the forces of
geographic inertia – for instance, via ever-greater fixed capital investment and the creation of a built infrastructure. The growth of productive
forces therefore acted as a barrier to rapid geographical restructuring:
'The more the forces of geographical inertia prevail, the deeper will the
aggregate crises of capitalism become and the more savage will
switching crises have to be to restore the disturbed equilibrium' (p.
429). In this fashion, he argued, regional crises built to create global
crises.

It is in this 'third cut' that many of the key statements of this text lie. *The
Limits to Capital* was significant particularly for its attempt to introduce
space at different scales to Marxist theorisations of crisis, and in so doing
it neatly emphasised the significance of an *historical-geographical materialism* to the understanding of social change. As an abstract theoretical
statement, of course, it could not be all-encompassing (the state, for
instance, was not systematically examined). More crucially, though,
Harvey deliberately and consciously separated much of the *theory* in his

account from the *history* and *geography* of capitalism. This was in one sense understandable: to say that the book was already a long and complex statement is certainly to understate the point. But as he remarked in the 'Afterword' (p. 451), it was important eventually to reconnect the book's theoretical endeavour with capitalism's social and spatial progress: 'in the final analysis, it is the unity which is important. The mutual development of theory and of historical and geographical reconstruction, all projected into the fires of political practice, forms the intellectual crucible out of which new strategies for the sane reconstruction of society can emerge'. It was in the need to find ways to *transcend* the limits to capital via the connection of theory, history and geography, therefore, that the book concluded.

From reformulation to fractious disintegrations?

Engagement with social theory had undoubtedly clarified many of the earlier conceptions of Marxist geographers. Geography had even begun to contribute, in a complementary sense, to social science more generally. But this broadening of the concern of human geography (partly due to the questions opened up in and through Marxist approaches), combined with the evolving status and state of Marxist theory, and the greater cross-fertilisation between social-science disciplines, had led also to a broader range of alternative analysis and criticism. In the process the claims raised by Marxists and geographers over the significance of class and place were exposed to a fierce, often internally divisive, and certainly increasingly fractious, re-evaluation. Established to a certain degree of legitimacy, as it were, in geographic environments, Marxists faced the twin problems of mapping a future course and responding to renewed external critique.

The tone of the latter was set in a guest editorial in the journal, *Society and Space*. (This had been founded in 1983 and was described by Richard Peet, one-time editor of *Antipode*, as representative of a school of thought associated with 'liberal analyses of various kinds by people who definitely did not share revolutionary ideas, people who regarded the word revolution as an anachronism' (Peet, 1985a, p. 4) – indicative perhaps of the emergent tensions.) In their editorial, Saunders and Williams (1986) identified what they saw as a new orthodoxy in urban studies, one 'embracing left Weberianism and the different varieties of Marxism, but excluding almost everything else' (p. 393). This, they argued, threatened to stifle the new perspectives that were necessary to analyse and understand contemporary social change. They questioned too the continued acceptance of class primacy in theoretical understanding: 'There is still a reluctance to accept that class may not be of primary significance in people's lives, or that there may be analytically distinct (though obviously empirically interrelated) bases of domination and conflict in society of which class is only one' (p. 399).

Harvey (1987b) responded to this attack in one of a series of editorials published the following year, in the form of a debate in the same journal. He identified three 'myths' in criticism of Marxist urban research and went on (p. 375) to reject strongly the claims of realism and postmodernism (see also Chapters 5 and 6):

Postmodernist philosophy tells us not only to accept but even to revel in the fragmentations and the cacophony of voices through which the dilemmas of the modern world are understood. It has us accepting the reifications and the partitionings, actually celebrating the fetishisms of locality, place or social pressure group. The rhetoric is dangerous for it avoids confronting the realities of political economy and the circumstances of global power.

To argue that place and particularity should have a stronger role in historical materialism, he maintained, need not entail abandoning universal statements and abstractions: 'I have long argued that historical materialism should be better thought of as historical-geographical materialism and that historical geography, particularly of capitalism, should be the primary subject of our theorising' (p. 376). And whilst political organisation had to be sensitive to the local, 'that does not mean that we bow down to parochialist politics and abandon the search for global solutions' (*ibid.*).

Clearly, the debate over urban studies, and the encounter with postmodernism more generally, touched upon fundamental *political* issues (and these are explored in Chapter 6 and the Epilogue). This was acknowledged by N. Smith (1987a); the strength of Harvey's comments, he argued, lay in the recognition of political implications at a time when 'politics are increasingly removed from the realm of legitimate discourse' (p. 378). 'Above all', he argued, 'let us debate the politics openly instead of relegating political agendas to a subtext' (p. 383). Much of this resonated strongly with the events of the late 1960s and early 1970s, and the calls then for relevance and political dialogue. In the late 1980s the particular context was most apparent from two debates: one over the significance and status of research into and on 'localities' and another (partly related) about the proclaimed emergence of a new production regime characterised by 'flexible specialisation'.

Massey (1983, 1984) had developed the notion of 'spatial divisions of labour' to focus upon the particularity of place and its significance to the process of capitalist restructuring. Such ideas came to underpin a range of research into and on places and regions, including a major programme funded by the UK Economic and Social Research Council known as the 'Changing Urban and Regional Systems' initiative (CURS) (see Cooke, 1986a; Robson, 1986). This rapidly became embroiled in a series of criticisms. N. Smith (1987b – see also the reply by Cooke, 1987b) argued that the CURS programme was primarily about the study of the specified localities in and of themselves, 'rather than an attempt to understand the dimensions of contemporary restructuring as revealed by the experience of these localities' (p. 407). This concern that 'locality research' should not go into small-scale empirical excess also had a deep *political* significance. As Cochrane (1987, p. 361) put it, the threat to political practice was that local concern led to 'economic microsurgery' – tinkering with the workings of the capitalist economy at a small scale but neglecting the system as a whole. N. Smith (1988, p. 151) expressed it succinctly: 'the danger for the left is that it becomes mesmerised by the new localism'. Cox and Mair (1989) made a conscious attempt to disentangle the growing controversy over localities, partly through the use of realist concepts (see Chapter 5). They concluded (p. 129), though in terms that reinforced Smith's critique, 'Locality under capitalism is irreducibly political. Locality is not just produced, it is struggled over. In consequence the political . . . has to be central to locality research'.

Figure 2.8   The 'localities' debate: culinary diversions and excursions

It is difficult to envisage a more fractious theoretical debate than that raised by the question of 'locality research' – one largely conducted (in print at least) through the pages of *Antipode*. Duncan and Savage (1989, p. 192), for instance, labelled 'locality' a 'dangerous concept' because through it, 'implicit notions of spatial determinism were smuggled back into urban and regional studies'. Their argument was that the concept was confused because it failed to provide an explanatory schema for dealing with spatial patterns (see also Duncan, 1989b). Whether this was in fact what locality research set out to do was another matter, but their contribution provoked a further heated set of comments. Cooke (1989b, p. 272), true to his role as co-ordinator of the CURS programme, replied as follows: 'the proof of the pudding will be in the eating. Locality can be seen to be a fascinating, complex concept of considerable value to geographical theory and empirical research'. Warde (1989, p. 274) picked up on this culinary analogy: 'Despite the increasing frequency with which it appears on the conceptual menu, "locality" is clearly proving somewhat indigestible; hence the heat in the kitchen concerning the best recipe for pudding'. He argued that whilst the dish should be made, its recipe required careful attention. He concluded (p. 280): 'Cooke's pudding, the heavily battered one made with self raising flour, is a bit shapeless. Duncan and Savage's alternative recipe seems to have overlooked the sugar and spice, a rather insipid dish of nouvelle cuisine proportions, scarcely pudding at all'. It was often difficult to disagree with Peet and Thrift's (1989, p. 22) verdict on this whole debate that it 'produced an enormous amount of heat, but it remains to be seen whether it will produce any light', adding perhaps that it did on occasion spark a little humour.

The questions raised of theory and practice by locality research involved increasingly bitter differences of opinion (see Figure 2.8), partly reverberating into and out of the encounter between postmodernism and Marxism (on postmodernism more generally, see Chapter 6: this account deals only with the specific confrontation between postmodernism and Marxism). As Lovering (1989) argued, countering the view expressed by Cooke (1989c), locality research need not necessarily be postmodernist (see, for instance, Beynon, Hudson and Sadler, 1991). The relation between postmodernism and Marxism was variously interpreted, too. To Graham (1988, p. 63), for instance, postmodernism was 'an emerging tradition within Marxism. Marxism is dead. Long live Marxisms'. Others (for instance, Soja, 1989) saw it as a logical outgrowth of and beyond Marxism. But to Harvey (1987a), postmodernism was a diversion from the central tasks of Marxism, nothing more than the 'cultural clothing of flexible accumulation' (p. 279). The goal remained 'to find a political response to the invariant and immutable truths of capitalism while responding to the particular forms of appearance that capitalism now exhibits' (p. 279). As he subsequently argued (see Harvey and Scott, 1989), the focus on empirical work and its justification via postmodernism's emphasis on difference represented a 'disengagement from explicit theoretical work rooted in political economy' in the face of neo-conservative political strategies, which had been 'an understandable but self-defeating retreat' (p. 222).

Even this explicitly theoretical work, though – in particular on the emergence of a new regime of flexible accumulation – found far from universal consensus. The debate over flexible production was largely couched in the language of the French regulationist school (see Aglietta, 1979; Lipietz, 1986). In particular there was an emphasis upon the new forms of production organisation that were held to be replacing the Fordist system (characterised by state-regulated manipulation of the balance between production and consumption to secure growth) that had broken down in the midst of world economic crisis in the 1970s. From different starting points, it was argued that a new era was dawning, variously described as post-Fordist, neo-Fordist or flexible production/specialisation/accumulation (see, for instance, Storper and Christopherson, 1987; Scott, 1988a, 1988b). This system rested on a fragmentation of the production process, as what previously seemed an inexorable trend towards larger, more vertically integrated units of production broke down under the weight of pressures towards vertical disintegration. In this model of economic development, it was also claimed, there was a tendency towards the spatial reconcentration of production as clusters of small firms emerged in new industrial districts: 'the vertical disintegration which lies behind flexible specialization creates powerful agglomeration tendencies at the regional level. Flexible specialization itself leads to the recomposition of the industrial complex, through a new form of horizontal integration of production capacity' (Storper and Christopherson, 1987, p. 115).

Whilst conceptually simple in outline, this idea that a whole new basis of production was emerging generated a heated debate (for a selection of responses, see Hudson, 1988; Pollert, 1988; Amin, 1989; Sayer, 1989b; Lovering, 1990). It was unclear in the first instance just how widespread a phonemenon 'flexible specialisation' was even supposed to be, let alone might be, despite the bold claims advanced for the generalisability of the model. The characterisation of the new regime as 'flexible' was also criticised in that it conflated system flexibility with labour flexibility; the latter most definitely not a liberating phenomenon from the point of view of the individual employee, but more often than not a prelude to further intensification of the production process. The notion of new regional production complexes also provoked reaction, in particular the charge that it rested upon a very partial and selective reading of the experience of parts of Italy, such as Emilia-Romagna (described in Brusco, 1982), where in fact the evidence pointed to considerable pressures on and difficulties for the companies and employees concerned. Finally, in its concern for abstract theory, the flexible specialisation model made some sweeping generalisations (which entailed a particular conception of the relationship between social and spatial change) that left little room for indeterminacy and human agency in production re-organisation. The grand theory of flexible production, in other words, systematically downplayed the role of social struggle and agency in the fashioning of structural change.

It was apparent that the path of debate within the Marxist tradition in the late 1980s had been a tumultuous and divisive one, in the midst and in the face of a growing variety of alternative analytical approaches. Having achieved a degree of incorporation and recognition within human geography, Marxist approaches themselves became the focus of trenchant criticisms. Some of the dangers to Marxist geography of such developments were recognised by Walker (1989) in an eloquent appeal to broaden the debate beyond narrow confines and re-establish a shared or common agenda. In cautionary words reminiscent of the earlier reaction by some to the rise of Marxist approaches in the early 1970s, he noted (p. 156) that

> Every wave of fashion has the force of good ideas and good intentions behind it. But there is more to the way people climb aboard each passing train of ideas than disinterested pursuit of scientific advance. Each wave opens up new horizons for young and agile minds looking for their own niche in a crowded occupation.

These new waves often repeated old errors, though, so that (p. 157) 'it is wise . . . not to fall in too quickly with every new enthusiasm that sweeps onto the academic shores'. Instead, he argued, it was 'important to acknowledge the collective nature of the left project' (from which he specifically excluded Saunders and Williams, whose 1986 editorial he condemned as 'petty' and 'scurrilous' – pp. 158, 159). Within this shared exercise, he went on (p. 158),

> the sources of discord could be very much greater than they are, if we were all to vent our spleen about every failing of reason or politics on the part of our colleagues. In this glass house of academe, it is altogether too easy to see the shortcomings in one's associates. Let us reason together.

Walker (*ibid.* pp. 160–1) concluded that, whilst there was much that Marxism did not answer, it was 'worth defending against glib dismissals by those who little understand its depths of analytic power or powerful link it forges between knowledge and change'.

Concluding remarks

One of the most significant features of Walker's clarion call for a rejuvenated Marxist geography is not so much the fact that he felt it necessary to make such a statement at all (although this is, of course, important), but rather the way in which it depended upon the notion of *collective* enterprise. For such a shared project had become, by the end of the 1980s, relatively unfashionable in a world where group (and in particular class) identity was apparently increasingly submerged in a whirl of social fragmentation, amidst the reassertion of individualism (at least by those in a position with the resources to do so). Much of this was subsumed in the proclaimed (by some) emergence of postmodernism as an alternative (dis)organising principle within society, with its celebration of difference and uniqueness and distrust of grand theory. In this climate the reforming ideals of Marxism, its notions of collective action and its generalised explanations seemed uneasy and uncomfortable. Other material conditions of social change led also to increasing uncertainty as previous monoliths seemed to vanish overnight – typified by the pace of political reform in Eastern Europe, with all the implications this carried for the state of global geopolitics. It was perhaps no surprise then that much Marxist work in geography in the 1980s had come to be characterised by uncertainty and friction, in the light of these twin challenges.

This situation was all the more difficult to respond to because of the problems experienced in actually pinning down and defining 'postmodernism' (see Chapter 6). Postmodernism's shadowy indeterminacy made it a most awkward perspective to grapple with, but one of its most important dimensions (from a Marxist viewpoint at least) was the *political*. For instance, Harvey's *The Condition of Postmodernity* (1989b, p. 116) argued partly that postmodernism's stress upon difference could take matters too far: 'It takes them beyond the point where any coherent politics are left, while that wing of it that seeks a shameless accommodation with the market puts it firmly in the track of an entrepreneurial culture that is the hallmark of reactionary nonconservatism'. Harvey went on to consider the nature of the changes taking place within twentieth-century capitalism and to highlight the political consequences of postmodernism, reasserting the primacy of class relations and the significance of space to political and geographical reform. Capital's increased sensitivity to the variation of place in the search for profit was one factor behind 'the production of fragmentation, insecurity and ephemeral uneven development within a highly unified global space economy' (*ibid.* p. 296). By contrast, labour's place-bound character left it particularly vulnerable, 'a part of the very fragmentation which a mobile capitalism . . . can feed upon' (p. 303). Only by simultaneously recognising and transcending the nature of place, he argued, could labour hope to build historical change.

Such remarks were doubly significant, not just in the context of the postmodernist challenge but also in terms of evolutions within Marxist geographic thought. Much of the debate that crystallised in the later 1980s on the nature of and relations between evidence and theory and, in particular, the proclaimed retreat from theory and support for place-bound political practice, which, for some, locality research represented, can only be understood in this setting. N. Smith (1989, p. 156) expressed this as follows: 'The shift towards empirical research was not only highly desirable but necessary if our knowledge of locational patterns, processes and transformations was to be advanced. Quite unnecessary and already debilitating is the fact that re-emphasis on the empirical was achieved by burying theory'. In many ways Marxist approaches to geography stood at the start of the 1990s at something of a crossroads. Many of the implications of a Marxist perspective were fully worked out and understood, but where future research paths lay was far less clear.

Whilst a great deal had been achieved through the incorporation of Marxist approaches to human geography, a lot remained to be done. Old advances required consolidation and new challenges had to be faced. Marxism as a body of theory is frequently criticised for what are represented as 'closed' assumptions and principles: a limited engagement with and accessibility to alternative ideas. The history of its adoption and subsequent reinforcement within geography largely gives the lie to such claims. If the rise of spatial science opened up geography to other ideas, then the Marxist challenge – especially in the early 1970s – cemented that dialogue in place and subsequently reinforced it with an openness to other concerns of social theory. In the process, it laid the basis for fresh criticisms of Marxism within geography. At the same time, however, in responding to the postmodernist denial of the possibility of grand theory, Marxist geography had neglected much of the practical thrust of the Marxist project: its connection with ways not just of viewing the world, but also of attempting to change it. Such a deficiency – both result of and condition for growing political uncertainty across the globe – was in danger of becoming a crippling one. And this is an important basis for evaluation of the Marxist contribution to human geography (if, admittedly, a difficult one), which should be put alongside the task of considering Marxist geography's intellectual contribution.

In part, though, we have anticipated the story in the rest of this volume. For much of the dialogue between Marxist approaches and geography more generally came in and through the *humanist* critique; and indeed it is not without significance that some of the most fruitful research areas being explored lay in the area of overlap between humanist and historical materialist approaches (see, for instance, Kobayashi and Mackenzie, 1989). The development of humanist approaches to human geography is therefore the focus of the following chapter.

3

'PEOPLING' HUMAN GEOGRAPHY AND THE DEVELOPMENT OF HUMANISTIC APPROACHES

Introduction

This chapter tackles the development of *humanist* – or, to use a term more commonly employed in this respect, *humanistic* – ways of approaching human geography. The heading of 'humanist(ic) geography' can be used to characterise a diversity of philosophies, methods and substantive studies present in the geographical literature, and it must be admitted that some of the authors discussed below might not even recognise this heading as one appropriate for their own work (and even a geographer commonly identified as a founder of humanistic geography has questioned the value of this designation: see Relph, 1977, 1981b). Moreover, it must be acknowledged that humanistic geography has developed in a more fragmented fashion over the past two decades than has been the case with the parallel development of Marxist geography, and in part this fragmentation has arisen because (whereas Marxist geography of whatever hue is ultimately rooted in the theoretical corpus of Marxism) humanistic geographers have ranged widely over a diversity of sometimes quite incompatible intellectual positions. Indeed, these positions have included foundational philosophies such as 'phenomenology' and 'existentialism'; methodologically inclined philosophies such as 'pragmatism'; philosophies of meaning such as 'hermeneutics'; theological arguments derived from the Christian Bible; psychological and psychoanalytical materials, notably those referred to as 'environmental psychology'; and also principles of substantive inquiry such as 'historical idealism' or 'interpretative sociology'. And this fragmentation has been made worse still by a strong fascination on the part of humanistic geographers with the specific ideas of particular 'great' humanist thinkers (see, for example, Seamon (1978) on Goethe; Cosgrove (1979) on Ruskin; Chappell (1981) on Tetsuro; Entrikin (1977, 1981) on Cassirer and Royce; Rose (1981) on Dilthey; Guelke (1974, 1982) on Collingwood; and Hasson (1984) on Buber). This appeal to a diversity of intellectual positions and to numerous different thinkers has added considerably to the flavour of humanistic geography, although the presence of so many different sources of inspiration warns us against implying that the marrying up of humanism and geography has been a straightforward, coherent and consensual process.

Attempting to make some sense of humanistic geography 'as a whole' is not going to be easy, then, but it is possible to identify a few principles shared by practically all of the practitioners we categorise under this umbrella heading. In the *Dictionary of Human Geography*, for instance, Derek Gregory (1986, p. 207) offers the following definition:

> An approach in human geography distinguished by the central and active role it gives to human awareness and human agency, human consciousness and creativity; at once an attempt at 'understanding meaning, value and [the] human significance of life events' (Buttimer, 1979) and 'an expansive view of what the human person is and can do' (Tuan, 1976).

It can hence be stated that the bottom line for humanistic geography lies in the objective of *bringing human beings in all of their complexity to the centre-stage of human geography*: and this is an objective pursued explicitly, in a sustained fashion and in conscious opposition to the curiously 'peopleless' character of much that had previously been passed off as 'human' geography. If we begin to look more closely, though, it soon becomes apparent that there are two fairly distinct components to this 'peopling' of human geography: one to do with recognising the humanity of the geographer and the other to do with recognising the humanity of the people the geographer studies. In the first respect, deliberate attention is paid to the geographer as an individual who enters very actively into his or her own research and writing; and in the second respect, deliberate stress is given to the crucial role played by human beings 'out there' in the 'real world' as people perceiving, interpreting and shaping the human geography of their surroundings.

In the first respect, disturbing philosophical and methodological issues are raised about the need to go beyond the conventional 'scientific' notion of the researcher as a simple toiler after objective truth, and about the need to recognise all of the complex a priori assumptions, values, hopes and fears researchers themselves cannot avoid stirring into the study of human situations. Having said this, we want to insist that the gulf between humanism and science is not so great as is sometimes implied, and that much of the philosophical material drawn upon by humanistic geographers is itself touched by a desire to parallel science, to refound science or to emulate science's transcendence of everyday human realities. In the second respect, a new package of 'things' to study – namely, human beings in all of their complexity – is opened up for incorporation at the heart of geographical inquiry and, in consequence, questions about how human beings come to know and to act in the world become of paramount concern. Whereas the complexity of the human beings who had the nerve to feature in their studies was once either ignored or assumed away by geographers, a quiet revolution has now made this complexity unavoidable in all but the most dyed-in-the-wool spatial scientific and Marxist geographies. In addition, it is vital to notice the concern shown here by humanistic geographers to *place* – to the myriad ways in which people 'relate' to the places around them where they work, rest and play – and to register that all manner of questions about people and their places have now spilled over from humanistic geography to energise human-geographical inquiry more generally. We ascribe considerable importance to this heightened awareness of the intimate emotional, practical, political and other attachments people usually possess with the places containing them, and as such we are prepared to talk about an emergent and innovative *geographical humanism* geographers can proudly introduce to the wider audiences of philosophy, social science and the humanities. We also want to claim that this geographical humanism is not particularly conformable with much of the prior humanist thought that supposedly inspired it, but that this lack of conformity should not lead us so much to criticise its practitioners – to quibble about their fidelity or otherwise to particular humanist philosophies – as to recognise the importance and the novelty of what they have achieved.

These introductory remarks give some indication of an argument that threads its way through this chapter, although it should also be noted that we have organised our discussion in a loosely chronological fashion that follows the fortunes of humanistic geography from its mixed beginnings to its present influential position. To be more specific, we begin by 'setting the scene' – by tracing the historical and theoretical roots to *humanist thought* – before proceeding to outline the 'prehistory' of humanistic geography in both early twentieth-century geographical writing and the 'behavioural' perspectives heralded during the late 1960s. What we discuss next is the emergence of a self-conscious humanistic geography from about 1970, and we identify within this approach an element of *critique* (a critical reaction against geography as spatial science) and also an element of *reconceptualisation* (a rethinking of basic geographical concepts). Finally, we describe the more recent elaboration of a geographical humanism that has brought the intimacy of everyday people–place relations to the forefront of both substantive inquiry and theoretical reflection throughout the discipline of human geography.

Humanist thought and humanistic geography

Writing in the early years of this century, F. C. S. Schiller (1907, p. 12) provided a cogent introduction to the tradition of humanist thought: 'Humanism is really in itself the simplest of philosophic view points: it is merely the perception that the philosophic problem concerns human beings striving to comprehend a world of human experience by the resources of human minds'. In other words, humanist thought insists that we take seriously the fact that the world is, to all intents and purposes, nothing but the sum of human experiences – the sum of human encounters with something we might call 'external reality' (the existence of which few humanists would deny) – and that we have no access to this world, no capability for knowing about this world, other than through 'the resources of human minds'. All of the intellectual materials available to us as we strive to make sense of the world around us are indelibly human attributes and products, and what this means in effect is that we are always both a 'barrier' between 'external reality' and our comprehension of that reality (to put things negatively) and a 'medium' through which comprehension of that reality is made possible at all (to put things more positively). The experiences that human beings have and the attendant workings of their minds are hence the fundamental 'data' of the humanist inquiry, and the humanist therefore 'insists on *leaving in* the whole luxuriance of individual minds' and on paying attention to 'the psychological wealth of every human mind and [to] the complexities of its interests, emotions, volitions, aspirations' (*ibid*. p. 13).

Schiller clearly calls for human beings in all their complexity ('warts and all') to be centre-stage in any intellectual reflection upon the world, even when this reflection is ostensibly directed towards the natural world of atoms and rain-forests, but it is important to note immediately that his arguments are more about installing humanity *in* the intellectual process than about conducting studies *of* humanity beyond the academy. His arguments are more in line with the first of the two components to humanistic geography identified above, then, and this is actually the case for the entire tradition of humanist thought: a tradition of thought that, for centuries, has been preoccupied with the problems and possibilities arising from the fact that reality is only knowable through the curious instrument of the human mind, and a tradition that has only turned incidentally to the study of particular human beings in particular worldly circumstances. It is perhaps to run ahead of ourselves, but what we will flag here is our own belief that much humanistic work being conducted within human geography actually reverses these priorities of humanist thought by concentrating more on studying humanity (and specifically on studying people in their places) than on 'philosophising' about the human factor in intellectual activity.

Schiller also indicates that humanist thought sees itself having a 'practical' dimension – and note that Schiller talks about both humanism and *pragmatism* (see Figure 3.6.) – and being concerned to mobilise 'the resources of the human mind' in the hope of improving the human condition: in the hope of making the world a better place for people to live in. He thereby claims that humanism aims for the 'complete satisfaction' of human beings, and that in so doing it must 'not cut itself loose from the real problems of life by making initial abstractions which are false' (*ibid.* p. 13). And to illustrate this point he mentions Protagoras, the scholar of classical antiquity whom he champions as the founder of humanism:

> If Protagoras had been a university professor, he would hardly have discovered humanism. . . . Fortunately he lived before universities had been invented to regulate, and quench, the thirst for knowledge; he had to earn his living by the voluntary gratitude for instructions which could justify themselves only in his pupils' lives; and so he had to be human and practical, and to take the chill of pedantry off his discourses.
>
> (*Ibid.* p. 15)

Schiller is not suggesting that humanism does away with the need for educational establishments, but he is worried that cloistered 'intellectualism' with no eye to the 'larger life' of humanity – with no eye to everyday problems of living, eating, sheltering, working and so on – leads to intellectual debate that is of little practical use to anybody. And yet this is not to advocate turning the academy over solely to the teaching of technical subjects such as bridge-building and needlework: rather, it is to recognise that 'our only hope of understanding knowledge, our only chance of keeping philosophy alive by nourishing it with the realities of life, lies in going back from Plato to Protagoras' (*ibid.* p. xiv). This insistence on nourishing intellectual activity with 'the realities of life', coupled with the reciprocal move whereby it is insisted that the goal of all intellectual activity is to improve these realities, is a definite feature of humanist thought (which is thereby criticised in error if it is said that humanism sanctions some sort of retreat from 'real life').

We have spelled out some of the basic ingredients of humanist thought using Schiller's words, but what we should emphasise now is indeed the deep historical roots of the humanist tradition. Schiller traces these roots back to the classical scholarship of Protagoras, but the more familiar claim is that humanism as a recognisable intellectual tradition took shape with the European 'Renaissance', a wide-ranging shift in intellectual, artistic and practical achievements that swept away the dust of Medieval times and heralded the dawn of more modern ones (and that had ramifications across a range of scholarly, cultural and scientific endeavours from philosophy to poetry and navigation; see Schmitt, 1988). At heart this Renaissance entailed nothing less than a profound transformation in what might glibly be termed the European 'world view': a transformation that affected the very way in which human beings conceived of themselves and of their role within the cosmic order, and a transformation that (to be more specific) gave humanity a much more important place in this cosmic order than had been the case in Medieval times:

> It is a truism to say that the Renaissance placed man [sic] at the centre of the cosmos, made man 'the measure of all things'. But there are few better phrases to capture the essence of the difference between a Medieval world picture centred upon God, wherein men and women held an important but decidedly inferior position in the great chain of being and in which their prime duty was to transcend the mortal world and their mortal bodies, and the new image of man as the central miracle of creation, the perfect measure of God and [of] God's creation, whose duty was, by exercise of reason – the supreme faculty – to know God through knowing and perfecting his creation. The distinguishing feature of the human individual was the capacity to reason, and human reason through the exercise of the arts revealed the consistency and pattern written by God in creation, a pattern that reflected the nature of the divine being himself.
>
> (Cosgrove, 1989a, p. 194; see also Cosgrove, 1984, 1985)

To put matters somewhat cryptically, the Renaissance effectively 'invented' the human subject as something indispensable to human thought and action: as something active in history, in the sense of 'making things happen', and as something capable of attaining a knowledge of both its external world (of natural phenomena, events, processes) and its internal world (of human nature, being and consciousness). And this manoeuvre restated the classical learning of Ancient Greece, and at the same time challenged the vision of the Medieval clerics in which humanity was portrayed as totally ineffectual and inherently ignorant in the face of God's supreme creation (and anyone who has read or seen Umberto Eco's *The Name of the Rose* should appreciate this point). The return to classical learning was in itself significant: and it should be emphasised that this return allowed the emergence of *the humanities* as distinctive fields of intellectual inquiry into the human condition, and that at this stage the humanities were supposed to include everything from 'History' and 'Geography' to 'Physics' and 'Mathematics'.

The fact that Renaissance humanism could embrace very different intellectual inquiries immediately indicates that what was involved here was no simple-minded pitting of the 'arts' against the 'sciences', even though today's advocacy of humanistic perspectives in disciplines such as human geography often amounts to little more than an opposition to scientific ways of thinking (see below). This is not to suggest that this opposition fails to identify important deficiencies in the procedures of conventional science, particularly when it comes to dealing with the ambiguities of human beings, but it is to indicate that it betrays a blindness to the roots of humanism that leads (we think) to misunderstanding of what certain humanist philosophies – and notably those of phenomenology and existentialism – are actually arguing. It is hence vital to recognise that one of the many faces of Renaissance humanism was what David Ley and Marwyn Samuels (1978b, pp. 4–7) call a *'scientific* humanism': a form of humanism typified by Galileo's mobilisation of the human senses (of sight, hearing and so on) in the context of a supposedly objective and empirical method (involving telescopes, observations and calculations) when 'proving' that the earth moves round the sun rather than vice versa. It is here that we encounter 'the exercise of reason' that occurred as human beings began to use the logic of the so-called 'scientific method' to uncover the workings of the cosmos, a cosmos assumed to be fundamentally ordered and law-governed in all of its parts (see Chapter 6), and it is here too that we encounter those philosophical reflections referred to as 'Cartesian rationalism' and indexed by the famous formula of 'I think, therefore I am'. Galileo (1564–1642) applied rational human thought in the guise of the scientific method to the understanding of the natural world, whereas Descartes (1596–1650) turned this thought back upon itself and thereby ushered in a crucial twist to the story of humanism:

> 'To think is to be' introduced a new dimension to the history of Western humanism. It introduced a fundamental *doubt* about existence itself, not to mention an existential doubt about one's own being. Thereafter, the central issue for Western science and philosophy was the proof of existence and the verification of one's own self.
>
> (Ley and Samuels, 1978b, p. 6)

What this meant was that Renaissance humanism spawned science as the pre-eminently 'human' tool for knowing the world, and that in so doing the capacity of the human mind to acquire knowledge about existence (and hence about itself) became a central problem for philosophical reflection.

However, it is important to note that Ley and Samuels draw a clear *distinction* between the basic tenets of Renaissance humanism and those of scientific humanism – and in effect they insist on casting 'proper' humanism and 'conventional' science in the oppositional roles mentioned above – and this distinction leads them to condemn as 'dehumanising' both the science of Galileo and the philosophy of Descartes. Ley and Samuels clearly want to secure an historical precedent for making humanity 'the measure' in academic inquiry, but in so doing they deny what others see as much more of a unity and complementarity between humanism *per se* and its scientific-philosophical offspring. Indeed, a commentator such as Edward Relph (1981b) sees 'scientism' (an obsession with the assumptions and potentials of conventional science) as an inevitable extension of the shortcomings within humanism itself, and therefore supposes that a humanistic geography committed to solving the problems of scientism – and thus setting itself up in opposition to spatial science – must at bottom be nothing but illogical. In this respect the suggestion is either that we begin to see humanistic geography as rather more compatible with spatial science (and other 'scientistic' approaches, including Marxist geography) than hitherto, or that we follow Relph in abandoning the project of humanistic geography in favour of something completely different again. Alternatively, many philosophers do not reckon Cartesian philosophies to be as 'dehumanising' as humanistic geographers normally imply, in part because these intellectual positions actually bring to the fore questions about the specificity of human beings – and here Descartes distinguished between the 'determined' qualities of *matter* (the world outside of human beings) and the 'free will' of *mind* (the world inside of human beings): a division that informed an early geographer such as Varenius (see Bowen, 1981, Chapter 2) – and in part because the very project that Ley and Samuels attribute to the Cartesians (that of probing issues of existence and selfhood) is one that rolls on into those modern 'philosophies of meaning' that so obviously influence the thinking of humanistic geographers such as Ley and Samuels themselves. Now we must acknowledge that the points being made here regarding the intersection of humanism and humanistic geography are quite subtle ones, but we feel it important to keep them in mind as we turn to a more 'narrative' description of how and in what forms humanistic geography has arisen and then changed.

The 'prehistory' of humanistic geography

Taking people seriously in 'early' human geography
Whereas there are few elements in the 'early' geographical literature (pre-1970) that anticipate the turn to a Marxist geography – although writing in 1949 S. W. Wooldridge did speculate that proclaiming geography to be Marxist would bring him 'extravagant adulation' from certain quarters (in Wooldridge, 1969b, p. 47) – it is possible to find numerous contributions to this literature that anticipate the turn to a humanistic geography. And searching out these earlier contributions is of more than mere antiquarian interest to today's humanistic geographers, given that well-known figures such as Anne Buttimer, David Ley and Edward Relph seek to glean 'lessons' from the past to inform their own theoretical and substantive inquiries in the present. What we might term the 'prehistory' of humanistic geography is a rich and diverse one, then, and it is impossible to do more here than gesture at this richness and diversity (and to suggest that considerable historiographic work remains to be done in recovering both the sophistication of the intellectual positions involved and the 'genealogy' of how these positions fed into one another and into more recent writings).

An initial point, however, is that at various times in the history of the discipline certain individuals have stressed the important role of geography as part of a thorough-going *humanist education* designed to give students a proper grounding in those intellectual areas – commonly referred to as the 'humanities' – concerned with establishing the more 'cultural' and 'moral' aspects of humanity. In part Cosgrove's interest in Renaissance humanism (see above) is inspired by a desire to recover the principles of humanist education, and then to rethink the purpose of teaching geography in the light of these principles:

> A humanist education incorporating the study of geography as the understanding of the richness and variety of the human world is individually and socially liberating. . . . A humane geography is a critical and relevant geography, one that can contribute to the very heart of a humanist education: a better knowledge and understanding of ourselves, others and the world we share.
>
> (Cosgrove, 1989a, p. 191, and Cosgrove, 1989b, p. 134)

And not dissimilar views to Cosgrove's have undoubtedly surfaced in the discipline before, as can be seen from David Stoddart's commentary on what he terms the 'humanising' of geography by late nineteenth-century scholars such as Peter Kropotkin (the 'Russian prince') and Elisee Reclus (Stoddart, 1986, Chapter 6). Indeed, in a famous essay entitled 'What geography ought to be' (1885) Kropotkin declared that the teaching of geography should help to overcome the 'jealousies and hatreds' which pitted nation against nation, class against class, and person against person: it should be 'a means of dissipating these prejudices and of creating other feelings more worthy of humanity' (Kropotkin, 1885, in Stoddart, 1986, p. 139). Another 'early' geographer strangely absent from most histories of the discipline is H. J. Fleure, the key figure in the Welsh 'school' of geography that centred on Aberystwyth during the first half of this century, and Fleure deserves mention here for his hope that geography would 'become more widely useful as an instrument of humanist education . . . and at the same time . . . a valuable influence in the enrichment of citizenship' (Fleure, 1918, in Bowen, 1976, p. 6). Fleure evidently disliked the British imperialism of Halford Mackinder, a contemporary of Fleure's who does feature in our textbook histories of the discipline, and it is revealing that he once contrasted Mackinder's geography with that of A. J. Herbertson – another 'early' geographer who linked geography to humanist education – by commending the latter 'for the effort of understanding peoples without the lust for power' (Fleure, 1965, p. 349).

But it was not just in talking about humanist education that 'early' geographers anticipated the emergence of humanistic geography, for when we consult various 'early' writings in the discipline we often find that these took people rather more seriously as 'objects' of geographical concern than is sometimes supposed. What we find more particularly is that appeal is often made by today's humanistic geographers to the *geographie humaine* that became associated with Paul Vidal de la Blache and his 'school' of French researchers during the first two or three decades of this century. Vidal developed an approach to geographical study that effectively challenged the older tradition of 'environmental determinism' (see Chapter 1) – a tradition that regarded all facets of human activity (from farming practices to political systems) as ultimately determined in character by the natural-environmental context (the mixture of climate, topography, hydrology, soils, flora and fauna) of the particular region under study – and in the process he proposed an alternative vision in which 'land and life' were supposed to influence one another in a genuinely two-way *mutual relationship*. In the Introduction to his posthumously published *Principles of Human Geography* (1926), for instance, he insisted on seeing in nature not 'definite, rigid boundaries' on human activity but 'a margin for the work of transformation or reparation which it is within [human] power to perform' (*ibid*. p. 24). More specifically, Vidal identified an ongoing 'dialogue' between natural environments and the human communities they supported – an ongoing interaction between *milieux* and *civilisations* – and he argued that the product of this dialogue was a human world full of different *genres de vie* ('lifestyles') distinctive to particular peoples living in particular places. In line with the full spirit of humanist thought as described above, he hence laid stress on the thought-and-action of human beings and on their ability to exert some independent control over their own destinies. Buttimer (1978, p. 61; see also Buttimer, 1971) puts matters like this when seeking guidelines for humanistic geography from Vidal's work:

Groups were seen to choose lifestyles according to their own insights, traditions and ambitions. They escaped from the tryanny of physical determinism by means of an idea: the idea they formed of their environment that impelled them to alter it. Within the realm of ideas, however, a tension could be observed between the creative and inventive force of human genius always tending to produce new patterns of work and dwelling, versus the conservative, 'sticky' force of habit that tended to resist change.

As this passage demonstrates, taking seriously human thought-and-action immediately introduces new sets of questions for the researcher (about, say, the tensions between innovation and tradition), but what we will simply underline here are the humanist overtones to Vidal's 'possibilism': his belief that the natural environment offers *possible* avenues for human development, the precise one chosen being very much a human decision (albeit one not always taken that consciously; see also Febvre, 1925). Moreover, in fostering a possibilist human geography Vidal effectively aligned himself with certain French philosophers who argued 'that determinism did not preclude the possibility of free will' (Berdoulay, 1978, p. 80; see also Berdoulay, 1976), and it is not difficult to see that the geographical debate here over determinism and possibilism interestingly prefigured the more recent geographical debate over structure and agency (see Chapter 4).

Antecedents of humanistic geography can also be found if we travel from France to North America, and notably if we consider the proposals for what J. K. Wright termed 'geosophy' in his 1946 presidential address to the Association of American Geographers (Wright, 1947). For many years previously Wright had been concerned with the geographical ideas possessed not by academic geographers but by all manner of people both great and humble, and in particular he studied the 'geographical lore' possessed by people living in 'the time of the Crusades' (in Medieval Europe):

By 'geographical lore' we mean what was known, believed and felt about the origins, present condition and distribution of the geographical elements of the earth .... It comprises theories of the creation of the earth, of its size, shape and movements, and of its relations to the heavenly bodies; of the zones of its atmosphere and of the varied physiographic features of air, water and land; finally, it comprises theories of the regions of the earth's surface .... Moreover, in addition to formulated beliefs, whether true or false, our definition of geographical lore covers [people's] spiritual and aesthetic attitude towards the various geographical facts, as revealed – often unconsciously – in descriptions of regions or of landscapes

(Wright, 1925, pp. 1–2)

Wright hence suggested that an important dimension to our labours as human geographers should entail asking about and striving to reconstruct the geographical lore – or, as David Lowenthal (1961) called it, the 'geographical epistemology' – contained in the minds of human groupings as diverse as the eskimos of Alaska and the chartered surveyors of Westminster. This is to push the geographical contents of human thoughts (but not necessarily actions) to the fore of our inquiries, although it might be argued that the humanism present in Wright's geosophy is relatively untheorised and as such avoids the philosophical difficulties that have tended to preoccupy humanistic geographers. This is not necessarily to criticise Wright's efforts, however, since in many ways his bypassing of overtly philosophical matters in the pursuit of other people's geographical lore has proved both a potent influence on later researchers – witness the 1976 collection entitled *Geographies of the Mind*

(Lowenthal and Bowden, 1976) – and a sort of 'prototype' for the most recent chapter of geographical humanism in the discipline's history (see below).

It must be reiterated that our observations in this section do little more than scratch the surface of 'humanistic geography *before* humanistic geography', but from what we have said it is possible, it is hoped, to appreciate certain hook-ups between the complexity of humanist thought as discussed in the previous section and the positions taken up by 'early' geographers such as Vidal, Wright and Fleure. To summarise, Vidal introduced a possibilist human geography in which the role of human thought-and-action in shaping both land and life (as opposed to the role of 'unthinking' natural-environmental forces) was brought to prominence; Wright proposed geosophical inquiries that took the researcher deep into the minds of diverse human groupings in pursuit of the specific geographical lores residing there; whilst scholars as diverse as Kropotkin, Reclus and Fleure exemplified a desire to arrive at a humanist education in which geography – and particularly 'regional geography' (see Chapter 1) – would be pivotal. Vidal, Wright and the others thus index an attractive humanist(ic) tendency in geography, a tendency well worth remembering and reassessing, and yet it was also a tendency that largely disappeared out of the door with the rise of a spatial science insensitive in so many ways to the humanity of both geographers themselves and people under study.

Spatial science and behavioural geography

As already outlined in Chapter 1, spatial science ushered in a very new approach to human geography from the beginning of the 1950s onwards. The numerous geographers who were caught up in the excitement of this development, initially in North America and then in Britain, were convinced that the way forward lay in casting geography as a fully-fledged science of spatial patterns and relationships: as a science like any other science, the only difference being that the purpose of geography was not to seek out laws of atomic motion, chemical reactions and so on, but to specify the fundamental *laws of spatial organisation* present in both natural landscapes and human activities in these landscapes. And, whereas 'early' geographers such as Fleure and Wooldridge (see the latter's 'Geographer as scientist' essay in Wooldridge, 1969b; see also Perry, 1990) had possessed a catholic sense of both humanism and science – had seen the two exercises as complementary, but had also recognised that there were many dimensions to 'being human', which could not be dealt with adequately in the narrow confines of a law-seeking scientific geogaphy – the 'Young Turks' of spatial science had no time for humanism, and instead supposed that a narrowly conceived science could help us to learn everything we wanted to know about human activity in geographical space. Indeed, what they wanted was a 'Newtonian' human geography: a human geography built upon a handful of foundational spatial laws equivalent to Newton's 'law of gravitational attraction', and it is telling that numerous spatial scientists tried to specify 'gravity models' that would explain and predict patterns of human interaction (migration flows, traffic flows, information flows) between settlements of differing sizes and distances from one another.

However, even from within the ranks of the spatial scientists a measure of unease quickly began to emerge about the inability of their cherished 'laws' and 'models' (notably their borrowings from Christaller, Von Thünen and Weber) to account adequately for observed patterns of location and movement within society. Some of them concluded that what was needed was a more 'probabilistic' or 'stochastic' approach to explanation, one in which chance or random factors were reckoned to play a significant role in 'distorting' *ideal* patterns of location and movement, but other researchers supposed that the only solution was a 'cognitive-behavioural' one designed to engage directly with the thoughts and actions of those human beings who fail to locate their factories in optimal locations or who fail to go shopping in the nearest supermarket (see Harvey, 1969a). Initially, then, there was an impulse from within spatial science to take seriously the more human attributes of human decision-makers (to recognise their inability to obtain 'perfect' information or to reason 'perfectly') and a handful of papers (notably Wolpert, 1964; Pred, 1967) signalled a promising way forward towards a form of 'behavioural location theory'. At the same time other geographers began to consider not so much the behavioural aspects of locating factories or visiting supermarkets, as the cognitive aspects of human beings perceiving and processing spatial-environmental stimuli (the layouts of buildings, roads, settlements and regions), and what this manoeuvre entailed was a shift in focus from human beings acting in the world to human beings simply reacting to it. Taken together these two developments effectively founded a new corpus of interest within the discipline that became known as *behavioural geography*, and in retrospect it can be claimed that behavioural geography acted as something of a 'bridge' leading from the 'peopleless' landscapes of spatial science through to the 'peopled' landscapes of humanistic geography.

To give a proper account of behavioural geography would require a volume in its own right, but what can be said here is that much of the work conducted under this banner restricted itself to a fairly narrow conception of how human beings think and act, and in so doing subscribed to versions of *behaviourism* – admittedly not always the simple 'stimulus–response' behaviourism that critics have suggested (see Figure 3.1) – rather than to other packages of psychological thought derived from (say) Freud, Jung or Gestalt theorists (see Gold, 1980, especially Chapter 2). Moreover, it appears that much behavioural geography has been conducted in a highly positivist fashion, with studies commonly employing statistical techniques such as 'factor analysis' and 'multiple regression analysis' to detect general patterns and law-like associations in large data sets (see, for instance, the studies written up in Golledge and Rushton, 1976), and as such behavioural geography should probably be considered more an 'outgrowth' of spatial science than a 'backlash' to it (Bunting and Guelke, 1979). In fact commentators such as Reginald Golledge and Helen Couclelis (Couclelis and Golledge, 1983; Golledge and Couclelis, 1984) are quite explicit about the continuing salience of positivism as a guiding philosophy for what they term 'analytical behavioural research in geography', although they also claim that certain 'limitations and constraints' of 'traditional positivism' have inevitably been overturned – and here they are thinking specifically of the difficulty that positivists encounter when confronted with the 'unobservable' phenomena of 'mind' – in the course of actually doing behavioural–geographical research. Nonetheless, it is wholly unsurprising that the move towards a self-conscious humanistic geography after about 1970 should have been accompanied by an increasing rejection of behavioural geography, and it might be argued that the initial stirrings of humanistic geography as critique (see below) were prompted as much by the *partial* treatment of people in behavioural geography as by their complete neglect in spatial science.

Figure 3.1　'Behavioural geography' and behaviourism

Behavioural geography can perhaps be described as a somewhat 'forgotten' perspective within human geography, despite the fact that it incorporates a number of concerns – the *perceptions* people hold of hazards and environments (see the important early statements in Lowenthal, 1967); the *mental maps* people have of the spaces and places around them (see Gould and White, 1974); the everyday *spatial preferences* people display in occupying or utilising certain spaces and places rather than others (see most of the essays in Golledge and Rushton, 1976) – which are still influential in much geographical thinking. One of the reasons for the lack of popularity endured by this perspective may lie in its allegiance to a 'school of psychology' called *behaviourism*, wherein tangible behaviours (whether those of human beings or animals) are viewed merely as 'stimulus–response (S–R) relationships in which particular responses [can] be attributed to given antecedent conditions' (Gold, 1980, p. 10). Such a perspective inevitably seems somewhat limiting and 'dehumanising' to many human geographers, particularly given that in extreme versions of behaviourism the cognitive processes of mind intervening between 'perceptual stimulus' and 'behavioural response' are almost totally ignored. Critics of behavioural geography commonly imply that behavioural geography simply apes this 'Pavlovian' or 'Skinnerian' behaviourism (see, for instance, Cullen, 1976; Ley, 1981a), but a fairer assessment would probably equate behavioural geography with Tolman's belief that a sensitivity to mind – and hence to 'the old questions of sensation and image, of feeling and emotion' – must still figure in the researcher's stimulus–response models and studies (Richards, 1974). It might be added that one or two writers have attempted to reconceptualise behavioural geography along non-behaviourist lines, but in practice the contents of the 'reflexive behavioural geography' proposed by the likes of Seamon (1984) are little different from those of what we now call 'humanistic geography'.

The emergence of humanistic geography

The element of critique

As explained in Chapter 2, a variety of social and political pressures external to the discipline began to exert an influence upon (and to suggest significant weaknesses within) geography conceived as spatial science. Whilst the most obvious consequence here was to encourage a human geography more attuned to the socio-spatial inequalities endemic to capitalist societies, it was also the case that some of the academics involved in rethinking the discipline at this time saw one dimension of its 'irrelevance' – its inability to speak usefully about 'real world' problems – to reside in the failure of spatial science to take seriously the complexity of human beings. As James Blaut (1984, p. 150) puts it when discussing the deeper tap-roots of the behavioural move in human geography during the late 1960s:

[This] was a time when science as a whole was under general attack for failing to solve human problems – worse, for failing even to notice human problems. A number of radical epistemologies and methodologies emerged from this critique. One of these, broadly described, was a demand that science return to the real, feeling, apperceiving, human being because here was the locus of suffering – in Vietnam, in Mississippi, in Haight-Ashbury and the college dorms, in ghettos and ill-designed public housing projects, but always in real, discrete, human bodies and minds.

In response to this demand many geographers did indeed embrace a behavioural perspective, then, but in some quarters it was felt necessary to extend the behavioural critique of spatial science by developing a much more trenchant critique: one that moved beyond behavioural tinkerings to challenge the entire philosophical framework in which spatial science had effectively (if not that self-consciously: see Chapter 1) been conducted. In short, a number of geographers set in train a whole new chapter in geographical thought: a new chapter in which the *positivistic* cast of spatial science was explicitly identified and was explicitly criticised through an appeal to an alternative package of humanist philosophies – notably *phenomenology* and *existentialism* – that carried with them an anti-positivist vision of appropriate questions to ask about human beings and of appropriate research procedures to employ in the process. And it was at this moment, as the 1960s shaded into the 1970s, that humanistic geography *per se* came into existence (although it had to wait until 1976 for Tuan to dream up the name) and began to establish itself as a potent source of philosophical *critique* directed at spatial science and its behavioural offshoots.

The starting point for this heightened attack on spatial science and its behavioural 'side-kick' was a basic concern about the curiously 'pallid' view of human beings – the horribly 'reductionist' view of man (*sic*) – being pedalled in these geographical enterprises. As J. Nicholas Entrikin (1976, p. 616) argued when reviewing the opening shots of humanistic geography,

> The humanist approach is defined by its proponents in geography and in other human sciences as a reaction against what they believe to be an overly objective, narrow, mechanistic and deterministic view of [the human being] presented in much of the contemporary research in the human sciences. Humanist geographers argue that their approach deserves the appellation 'humanistic' in that they study the aspects of [people] which are most distinctively 'human': meaning, value, goals and purposes.

It almost goes without saying that the geographers referenced here were unhappy about the tendency of spatial science to treat people as little more than dots on a map, statistics on a graph or numbers in an equation, since the impression being conveyed was of human beings 'whizzing' around in space – travelling from place X to place Y; shopping in centre X rather than in centre Y; selling produce at market X rather than at market Y – in a fashion little different from the 'behaviour' of stones on a slope, particles in a river or atoms in a gas. Indeed, it was complained that such exercises 'objectified', 'reified' or (in Olsson's more memorable term) 'thingified' the people under study, and that in so doing they effectively converted human beings into 'dehumanised' entities drained of the very 'stuff' (the meanings, values and so on) that made humans into humans as opposed to other things living or non-living. Various strands of philosophical argument were brought to bear in this connection, but

Figure 3.2   Leonard Guelke's 'idealist human geography'

Idealism, 'in its philosophical sense, is the view that mind and spiritual values are fundamental in the world as a whole' (Acton, 1967, p. 110), and as such it is opposed to *materialist* doctrines that 'give to matter a primary position and accord to mind (or spirit) a secondary, dependent, reality or even none at all' (Campbell, 1967, p. 179). This means that both humanist thought in general and humanistic geography in particular can be described as 'idealist', but to make this identification is to do little other than signal their philosophical distance from positions such as *positivism* (see Chapter 1), *Marxism* or *historical materialism* (see Chapter 2) and certain versions of *realism* (see Chapter 5). A much more specific manoeuvre has been made within human geography by Guelke, however, since in various essays and texts (notably Guelke, 1974, 1981, 1982, 1989a; see the much earlier statement of Lowther, 1959), Guelke has drawn upon the 'historical idealism' of Collingwood – a practising historian – in order to argue for a human geography that sees its role as one of *rethinking the thoughts behind the actions* (the 'insides' beneath the 'outsides') of 'human events' with tangible environmental-landscape impacts. In part Guelke is seeking to overcome the deficiencies of a behavioural geography that has abandoned its concern with human beings *acting* in the world, and which thereby fails to illuminate 'overt behavior of geographic significance' (Bunting and Guelke, 1979; see also Philo, 1989b), and in part he hopes to overcome the 'theoreticism' of much contemporary human geography by claiming that we need no theories of our own (whether these be Marxist, humanist, structurationist or whatever) because the only theories that matter are those possessed by people under study (and which shape their decisions to settle in particular environments, lay out their land in particular ways, locate their factories in particular places, and so on). His 'idealist human geography' has been heavily criticised (see Gregory, 1976; Watts and Watts, 1978; Harrison and Livingstone, 1979; Curry, 1982a), but it perhaps deserves credit for formalising an approach to explanation that has long been present in human geography – and especially in historical geography – and for complementing (notably in Guelke, 1989a) the heightened sensitivity to the ideas of given peoples in given places urged upon us by what we are calling geographical humanism (see later).

---

of considerable importance was Gunnar Olsson's demonstration that the 'language of social science' – with its insistence upon framing studies in the terms of *certainty*; in the vocabulary of law-like statements where $X = Y$ – cannot cope with the more *ambiguous* 'language of human action' in which most people think and act as they strive to 'get by' in everyday life (see the essays collected in Olsson, 1980; see also Philo, 1984). At the same time numerous geographers with humanistic leanings discussed the limitations that arise when thinking about human decision-makers using either the 'rational economic man' models of spatial science (Wallace, 1978) or the various 'stimulus–

response' models of behavioural geography (Cullen, 1976; Ley, 1981a), and in both instances objections have been raised to the presentation of individuals as 'mechanically determined automata' capable of no creative input to their location decisions or spatial behaviour. (See also the influential *idealist* alternative in human geography that Guelke proposes as a way of investigating 'creative' human influences on the geography of the world – see Figure 3.2).

It is perhaps a little curious, though, to find that the early texts of humanistic geography rarely articulated their disquiet with spatial science in quite these terms, which actually owe much to those more recent commentaries of geographers who see in spatial science a blindness to human agency occasioned as much by its neglect of 'social theory' as by its inherent philosophical shortcomings (and in this connection 'structurationist' appraisals and reworkings of human agency in social theory became vital; see Chapter 4). In fact, we cannot help feeling that the early texts of humanistic geography never quite did what logically they might have been expected to do: in that, rather than developing a detailed critique of drawbacks in how spatial science dealt with the people it studied, these texts spent most of their time reflecting upon drawbacks in the sort of knowledge the spatial scientist as researcher was able to glean from his or her studies. What is hence apparent from these texts is the central claim that the researcher should not be seen as an individual whose humanity stands outside of the research process: as an individual who is nothing but a vessel for taking in information, processing it and then arriving at conclusions. This image is central to how conventional science is supposed to proceed, and it is thereby a central tenet of positivist epistemology (see Hill, 1981, pp. 51–2), but it is one that can be easily criticised by pointing to both the sociology of how knowledge is actually 'produced' by particular communities of scholars (see Kuhn, 1970; Lakatos, 1970) and the philosophical issues involved in how people can ever generate insight into the world around them. As already explained, humanist thought is ultimately anchored in precisely these latter philosophical issues – in the fact that knowledge can never be anything other than rooted in 'the resources of human minds' – and it is hence unsurprising that geographers who were becoming increasingly dissatisfied with the alleged objectivity of scientific-positivistic research procedures looked increasingly to humanist thought for alternative philosophies of knowledge (which is why Entrikin (1976) describes the rise of humanistic geography as at bottom an exercise in criticising the positivism of 'scientific geography').

A key point made by the humanistic geographers was that all manner of human *values* are always bound up in the efforts of the researcher to know about their objects of research (whether these be natural or human), and Anne Buttimer in particular spelled out the character of these values permeating the geographical research process (Buttimer, 1974; and see the appended 'commentaries' on this paper). Scientific-positivistic research inevitably ignores this very 'human' complication to what is being attempted: 'In the scientific or "naturalistic" mode of knowing [the researcher] may become so engrossed in the objects of his [*sic*] concern that he overlooks himself and the perspectives he brings to the study of these objects' (Buttimer, 1976, p. 279). There is of course some overlap here with the more radical critique of conventional science for failing to recognise its own biases – often biases that predispose it to conduct research and to arrive at findings supportive of the social status quo (see, for instance, Harvey, 1972, 1974b) – and more particularly it fuses with

Figure 3.3   Phenomenology

The 'father' of phenomenology is usually taken to be Husserl (1859–1938), and – interestingly enough – it is claimed that Husserl built on the earlier work of Brentano, who 'sought to work out the logical geography of mental concepts as a necessary preliminary to any empirical psychology' (Findlay, 1960, p. 188). Parallelling this pursuit of philosophy as a 'rigorous science' of mind, Husserl hoped to establish a secure framework of concepts with which to 'ground' all of the other sciences, and as such his phenomenology was designed 'to disclose the world as it shows itself *before* scientific inquiry, as that which is pre-given and presupposed by the sciences' (Pickles, 1985, p. 3). He was extremely critical of the scientific-positivistic attitude (what he called the 'natural attitude') predominant around the turn of the century, principally because of its assumption – an assumption with 'objectivist' and 'empiricist' overtones – that the *objects* of the world can be known to human *subjects* in an entirely unproblematic fashion. He duly strove to overcome this subject–object dualism, and in so doing he provided what has been termed 'a descriptive philosophy of experience' (Farber, 1960, p. 292) concerned to elucidate the way in which objects must always be understood as objects *for* human subjects: as objects that human subjects *experience* (or gain *consciousness* of), and as objects towards which human beings always possess *intentions* of using or interacting with (however *un-* or *sub-*consciously). Husserlian phenomenologists hence suppose the major philosophical task to be one of getting 'back to the things themselves', not in a scientific-positivistic sense of assuming that the objects of the world simply 'speak' their truths to us, but in the much more profound sense of stripping away all of the superficial clutter of human minds deposited there by both our scientific and our common-sense understandings of the world – and this process of stripping away or of 'bracketing out' is referred to as the *epoché* or 'phenomenological reduction' (Entrikin, 1976; Johnson, 1983) – and of then being in a position to discover the true *essences* of objects that reside in our (in *everybody's*) deepest intentional relationship with these objects. It is further suggested that phenomenological reflection upon this place of deepest communion between objects and subjects, a place Husserl sometimes refers to as the *lifeworld* or 'universal horizon' *common to all humanity*, will allow researchers to refound all of the sciences from physics to geography on the basis of a true appreciation of the world's objects in relation to the world's subjects. It should hence be clear that Husserlian phenomenology is actually a very strict, demanding and yet (to many of us) strangely 'metaphysical' project, and as such it is hardly surprising that various philosophers after Husserl have sought to soften or to recast the outlines of this project in a manner that might prove more helpful to other areas of intellectual activity. To be more specific, the likes of Merleau-Ponty and Schutz have posited alternative phenomenologies – *existential* or *constitutive phenomenologies* – that preserve something of the spirit of Husserl's philosophy, but that talk less in terms of transcending the everyday and more in terms of studying precisely those everyday meanings etched into the *lifeworlds* of particular peoples, societies or cultures. And this manoeuvre has proved attractive to geographers such as Ley, who want to mobilise humanistic arguments whilst still studying everyday social geographies (see Ley, 1977, 1978, 1981a; and see also the commentaries in Entrikin, 1976; Gregory, 1978, Chapter 4). Interestingly too, some of the alternative phenomenologies – and notably those that inform the contributors to *Dwelling, Place and Environment* (Seamon and Mugerauer, 1985a) – have turned back towards psychology, but have avoided Husserl's transcendentalism in favour of 'concern, openness and clear seeing' when confronted with 'human experience and meaning *as they are lived*' (Seamon, 1987; and particularly important in this respect has been the so-called Duquesene 'school' of phenomenological psychology).

the demands for theoretical work in human geography to become more critical of its own 'knowledge-constitutive interests' (to use Habermas's term; see Gregory, 1978, especially Chapters 2 and 5). But for the humanistic geographers matters were perceived less in terms of how (say) the 'needs' of capitalism shape research in progress, and more in terms of how researchers 'get in the way' of their own studies precisely because they are human subjects whose personal 'subjectivities' blind them to the real nature of the objects they are studying. And the vocabulary employed here, as coupled with Buttimer's complaint about 'naturalistic' approaches to knowledge, indicates that the line of reasoning being pursued is indeed a philosophical one and, moreover, is a philosophical one rooted in the difficult Continental philosophy of *phenomenology*.

In Figure 3.3 we provide a few summary remarks about this philosophy, but what it is important to notice here is that the 'founding father' of phenomenology, Edmund Husserl, framed his inquiries very much as a reflection upon the dominant 'natural attitude' in which scientific-positivistic researchers ignored the question of their *own* involvement in the research process. In his well-known 'Philosophy and the crisis of European man' (1936 in Husserl, 1965) essay, he identified a 'sickness' in European society: a sickness that stemmed from the extension of the natural attitude to the domain of philosophy, and that carried with it an insistence that 'spirit' (the inner world of human being) should be studied with *exactly* the same techniques as those employed by the scientists of nature (the outer world beyond human being). But Husserl's argument here must be understood very carefully, for his objection was not to the natural attitude *per se* – far from it, for he admired its 'greatness' – but to the failure of humanist thinkers to develop philosophy as an equally 'rigorous' examination, albeit one requiring its own *distinctive* methods, capable of disentangling the 'truths' of humanity from the chaos of individual and collective psychologies:

> Blinded by naturalism (no matter how much they themselves may verbally oppose it), the practitioners of humanistic science have completely neglected even to pose the problem of a universal and pure science of the spirit and to seek a theory of the essence of spirit as spirit, a theory that pursues what is unconditionally universal in the spiritual order with its own elements and its own laws. Yet this last should be done with a view to gaining thereby scientific explanations in an absolutely conclusive sense.
>
> (Husserl, 1936, in 1965, pp. 154–5)

Husserl hence talked about 'philosophy as rigorous science' being the goal of proper humanist thought, and as such he clearly wanted humanist philosophy to *parallel* science in searching for 'universal and absolute' (or 'transcendental') 'elements and laws' governing the workings of the human spirit (and in this sense he must be seen very much as an heir of the 'Cartesian rationalism' described above; see also Fell, 1979, Chapter 1). Husserl's phenomenology was evidently no warrant for simply saying that human subjects are extremely complicated and therefore basically very different from one another in their respective psychological make-ups: instead, it was definitely concerned to establish what is common in these make-ups, in part as the conclusion of 'philosophical science' itself and in part as an input to the deeper conceptual bases from which other sciences (both natural and human) could proceed. (And note that seen in this light phenomenology is criticised in error if it is portrayed as seeking out entirely personal and therefore idiosyncratic 'phenomenological world-pictures'; see Mercer and Powell, 1972; Billinge, 1977.)

Husserl thus conceived of his project as both a 'philosophical science' and a radical reflection on the foundations of all other sciences, and it is in this latter respect – where a sort of 'blueprint' is laid down regarding how researchers as human subjects are supposed to disclose the true *essences* of objects under study (see Figure 3.3) – that humanistic geographers drew upon his phenomenology when considering how the humanity of the researcher should be (re)introduced into the geographical research process. Husserl argued that the positivistic 'empirical sciences' happily deal with the 'things' of the world, capably identifying laws that seemingly rule their disposition and behaviour, but that the success or failure of these sciences depends upon an a priori philosophical conceptualisation of essences that are viewed simultaneously as *explanations*:

> Positive science simply is not interested in what 'things' are. However, when the science in question is philosophy, when the object under investigation is being – whether the being of a 'thing', a 'state of affairs', a process, an event, a social reality or a culture – the investigator cannot be satisfied with a knowledge only practically valid [a knowledge sufficient for the practice of identifying laws of how 'things' relate one to another]. It is the philosopher's task to penetrate to the deeper validity rooted in the very essence of the object under investigation, which essence, of course, is also the ultimate explanation for the way things act – because they are what they are. . . . [F]rom Husserl's point of view, the essences of things will be the truth of knowledge about them.
>
> (Lauer, 1965, p. 45)

Or, as one geographer (Christensen, 1982, p. 51) summarises Husserl's position, every science (economics, sociology) 'possesses its own domain of investigation – its "region" – but the scientists cannot clarify this domain for themselves using the methods and logic of empirical science', and thereby should draw upon phenomenological methods 'in order to reveal the essential structures of the entities under study by the discipline' (*ibid.*). Associated with and in a sense 'grounding' each of these disciplines the phenomenologist is duly required to uncover a 'region' or 'regional ontology' specifying the essences of their basic objects of study, such as the essence of 'economy' and 'money' for the economist or the essence of 'society' and 'state' for the sociologist. And, to re-emphasise the broader case being made here, these founding essences of the sciences must be understood as residing in neither the objects under study nor the human subjects doing the study, but in the 'primordial relationship' that exists between objects and subjects as equivalent 'beings in, alongside and toward the world' (Pickles, 1985, p. 17).

These are undoubtedly complex claims – and they do seem quite alien to the way in which most of us have been 'trained' to think about science (and about the relationship between science and philosophy) – but what we need to recognise in this context is that humanistic geographers such as Anne Buttimer (1976, 1980), Edward Relph (1970, 1977, 1981a, 1985), David Seamon (1978, 1979, 1980, 1984, 1985b (with Mugerauer), 1987, 1989) and Yi-Fu Tuan (1971, 1974a, 1974b, 1976, 1977, 1979) have all turned to versions of phenomenology in an attempt to delimit an approach to geographical objects that recognises *philosophically* the centrality of the human subject not just in 'accessing' but in fundamentally 'constituting' the knowledge to be had concerning these objects. Their turn to phenomenology was certainly designed to critique (and, in the more charitable accounts, to complement; see Walmsley, 1974) scientific-positivistic spatial science, given that this approach naïvely reckoned itself able to know the 'truth' of geographical objects without any sort of self-reflection, but what must also be recognised is that the process of turning to phenomenology immediately led the humanistic geographers away from critique – and they were never as hung up on attacking other versions of human geography as is often supposed – and into the more 'constructive' project of elaborating an appropriate 'regional ontology' from which geographical inquiry could be (re)launched. The result was the practice of a *phenomenological geography*: a practice scientifically minded geographers could not recognise as such and a practice other geographers such as Kathleen Christensen (1982) and John Pickles (1985) could suggest was not phenomenologically rigorous enough, but a practice nonetheless.

The element of reconceptualisation

This turn to the practice of a phenomenological geography took various forms, and – as we shall see – it was in this respect that the humanistic geographers began to combine their phenomenology with varying degrees of commitment to the related (and equally difficult) Continental philosophy of *existentialism* (see Figure 3.4). Yi-Fu Tuan's 1971 paper on 'Geography, phenomenology and the study of human nature' provides a good indication of what was involved in this rethinking of basic geographical concepts, since here Tuan adopted a phenomenological approach designed to tease out the 'essences' of certain 'geographical concerns' residing in the deepest psychological, emotional and existential attachments that all human beings hold for the spaces, places and environments encircling them (and in the process Tuan (p. 181) also aimed to make some larger claims about the centrality of the world's geography to human nature itself):

> The theme of the paper is 'geography as the mirror for man' [*sic*]. The approach is phenomenological: for my purpose I take this term to mean a philosophical perspective, one which suspends, in so far as this is possible, the presuppositions and method of official science in order to describe the world as the world of intentionality and meaning. Phenomenology is concerned with essences: what, for example, is the essence of man, space or experience? . . . Geography . . . is 'organized knowledge of the earth as the world of man'. It needs only a slight recasting and expansion to cover my own position: knowledge of the earth elucidates the world of man; the root meaning of 'world' (*wer*) is in fact man; to know the world is to know oneself.

Figure 3.4   Existentialism

Existentialism as a philosophy is perhaps most closely identified with Sartre (1905–80), who argued that what existential thinkers have in common is the belief that '*existence* comes before *essence*' (Sartre, 1948, p. 26). For some thinkers this view signals the fact that things simply exist – that they simply are present in the world – and that beyond this observation there is really very little of value the observer, and particularly the researcher with his or her analyses and interpretations, can say about these things. For other thinkers this view emphasises the 'subjectivity' of human beings, the only *beings* in the world to possess 'will and consciousness', and insists that human beings are 'free' to choose the 'nature' of their existence and to ascribe it with *meaning*. As Sartre (*ibid.* p. 28) put it, 'man [*sic*] is nothing else but that which he makes of himself'; or, as Quinton (1988, p. 297) puts it, the human being 'is a self-creating being who is not initially endowed with a character and goals but must choose them by acts of pure decision, existential "leaps" analogous to that seen by Kierkegaard in the reason-transcending decision to believe in God'. At the centre of existential thought questions arise about the relationship between human beings and non-human things, and it is often claimed that human beings are fundamentally 'alienated' from the world of things – which are just so different from them – and that people are therefore engaged in a constant striving to 'make things meaningful' so as to fill the 'existential void' (the complete lack of meaning) at the heart of the human condition. These claims are sometimes connected up to Marxist arguments about how the 'capitalist mode of production' alienates humanity from nature, and about how the tendency towards ever-expanding commodity production under capitalism is fuelled by people's desire to fill the existential void with more and more material possessions (for discussions in the geographical literature, see Evans, 1978; Cullen and Knox, 1982). Heidegger (1890–1976) adopted the procedures of Husserlian phenomenology in seeking to understand the founding 'structures' of 'being-in-the-world', meanwhile, and here he argued that *Being* (or *Dasein*) is anchored in the 'primordial' ('taken-for-granted' or 'little-reflected-upon') spatial and temporal relationships people have with both the things alongside them in the world and the inescapable fact of their own finitude or mortality.

Tuan hence envisaged the discipline of geography and the philosophy of phenomenology as closely bound up one with the other, the former being given a more secure epistemological foundation through phenomenological reflection as well as offering materials whose philosophical inspection allows us better to appreciate the existential dimensions to human 'being-in-the-world' (a favourite existential notion).

To be more specific, under the heading of an 'existential response' to geographical concerns Tuan looks not for the *order* of dots, lines and hexagons but for the *meaning* a space, a place or an environment possesses because 'it is a sign to something beyond itself, to its own past and future, and to other objects'. Or, as Tuan (*ibid.* p. 184) also puts it: 'The geographer's concerns can now be restated. Under "environmentalism" he [*sic*] seeks meaning in order – and finds a largely determined, timeless and tidy world; under "existentialism" he seeks meaning in the landscape, as he would in literature, because it is a repository of human striving'. In the heart of his paper Tuan illustrates this claim by suggesting that there are 'essential' ways in which all people inscribe meaning in – and also derive meaning from – the world's geography, and here he begins by sketching out the 'egocentric' and 'ethnocentric' tendencies that emerge as individuals and groups endeavour 'to organise the world around themselves': as individuals interpret their personal relationships in terms of how close or distant they feel from loved ones and enemies, or as groups such as Greenland Eskimos develop cosmologies that are circular, symmetrical and with their own small settlements pictured as lying at the centre of everything. Similarly, Tuan discusses the basic 'front–back' asymmetry of the human body and argues that this asymmetry has a dramatic impact upon the human organ- isation of space, from the conscious delimitation of front and back rooms in buildings – these rooms being allotted different functions, the front ones being more 'public' (hallways, reception rooms) and the back ones being more 'pri- vate' (bedrooms, toilets) – to the less conscious delimitation of front (primary, 'spectacular') and back (secondary, 'tacky') routes into cities both ancient and modern. And again, he addresses the 'home–journey' distinction, explaining that people have a universal tendency to identify particular places as 'home bases', which are then equated with nurturing, rest, security and refuge, and which are distinguished from places more distant both physically and emo- tionally and that can only be reached following the hard work, insecurity and even danger of the 'voyage'.

Tuan considers other basic geographical concepts both in this paper and in other publications, and note that, in his 1976 paper on 'Humanistic geogra- phy', he declares the very purpose of this approach to be the description of 'how geographical activities and phenomena reveal the quality of human awareness' (p. 267). But it is perhaps in a later paper by Edward Relph that we can find the most explicit statement regarding the role of phenomenological geography – or, to use the slightly different but indicative phrase of Relph's, the 'phenomenology *of* geography' – in reconceptualising the wider discipline, since here he argues that 'academic geography' is a codification of the more everyday 'geographical experience' everybody has of the world around them (of their 'being-in-the-world'):

> The experiences of places, spaces and landscapes in which academic geography origi- nates are a fundamental part of everyone's experience, and geography has no exclusive claim to them. Indeed, one of the first aims of a phenomenology of geography should be to retrieve these experiences from the academic netherworld and to return them to everyone by reawakening a sense of wonder about the earth and its places. To do this can nevertheless provide a source of vitality and meaning for geography by casting it in its original light, where 'original' has the dual meaning of 'first' and 'new'.
>
> (Relph, 1985, p. 16).

Relph hence imagines a kind of 'slippage' occurring from our little-reflected- upon experience of the everyday 'ready-to-hand context or background' of the world in its more geographical manifestations – the woods we walk through;

the hills in the distance; the buildings in the high street; the railings around the school – through to the researching and the writing of human geography by academics. The objective for Relph is thus to recover the true character of our everyday geographical experiences, and in particular to recover what he believes to be a dual sense of 'marvelling' and 'concern' (or 'care') embedded in these experiences, as a prelude to reshaping an academic geography in which 'abstract technical thinking has begun to submerge geographical experience either by making (it) seem relatively trivial or simply by obscuring it with generalizations' (*ibid*. p. 28). More specifically, what Relph does is to examine four basic geographical concepts – those of region, landscape, space and place – that are not just concepts for academic geography, but are also 'the contexts and subjects of geographical experience, and in a different aspect again . . . are parts of being-in-the-world' (*ibid*. p. 21). We cannot pursue the details of Relph's inquiry here, but we can perhaps mention his account of 'place experiences' as 'constructed in our memories and affections through repeated encounters and complex associations', and as giving rise in almost all of us to a feeling for place as 'an origin: it is where one knows others and is known to others; it is where one comes from and it is one's own' (pp. 26–7). This is not to say that we will always possess positive affections for the places in which we are born, raised, work, rest and play; indeed we may sometimes possess strong aversions to them ('topophobias' as opposed to 'topophilias', to use Tuan's terminology) – but it is to identify a manner of thinking about place that is very alien to the spatial scientific treatment of place as a location (an (x,y) coordinate in space) or as a simple geometric assemblage of fields and farms, freeways and factories. (We will return presently to this important reconceptualisation of place.)

Relph's paper on 'the phenomenological origins of geography' is contained in a collection of essays entitled *Dwelling, Place and Environment* (Seamon and Mugerauer, 1985a, 1989), and taken as a whole this volume stands as an impressive testament to the accomplishments of a phenomenological geography concerned to rethink the deepest well-springs of human-environmental relations. Throughout the volume basic geographical concepts are reconsidered, and in effect an attempt is made to enrich these concepts by elucidating previously 'taken-for-granted' ways in which human beings interact with the natural, human and 'sacred' spaces all around them. Furthermore, various contributions are clearly inspired by David Seamon's work (1978, 1979, 1980, 1984) on what he calls 'the geography of the lifeworld' – a geography sensitive to the prereflective intentionality of the human body displayed in such simple but complex actions as walking, lifting food up to the mouth, turning on the television and so on – and in this connection we encounter a rethinking of the ground for geographical inquiry in which human 'apprehensions' (human perceptions, feelings and intuitions, however 'vaguely' sensed) are so bypassed that to talk of this rethinking as an exercise in 'humanistic' geography might actually be questionable! This is not to suggest that Seamon's version of phenomenological geography is without interest, however, and his notions of 'body ballets' and 'place ballets' – of 'habitual body behaviours' organised into 'time-space routines' through which *the bodies themselves* acquire a distinctive 'sense of place' (see below) – may provide an intriguing new way of studying the human geography of phenomena such as towns and markets (Seamon, 1979, 1980; Seamon and Nordin, 1980). And leading from a similar basis in the 'body-subject' phenomenology of Maurice Merleau-Ponty, Miriam Helen Hill (1985) finds that the 'body-world communion' of blind people hints at a 'holistic environmental knowing' in which touching, smelling and hearing (as well as seeing) all play a part in allowing one to 'read' the geography of one's immediate surroundings. What is at stake here is not solely the body's pre-reflective intentionality, though, for both Hill and Seamon point to ways in which this level of interaction with our surroundings feeds into a more 'conscious awareness' of how we all 'encounter' the world through the medium of our bodies, and how in the process we attain an awareness that – certainly if cultivated – can be absolutely 'crucial to who and what we are' (Seamon, pers. comm.).

But what is perhaps most noticeable about the writings of Relph, Tuan and the other phenomenological geographers mentioned here is that their reconceptualisation of basic geographical concepts almost inevitably takes them towards the existential arguments of philosophers such as Martin Buber, Martin Heidegger and Jean-Paul Sartre (see Figure 3.4). Indeed, in striving to rethink basic geographical concepts it is surely unsurprising that the humanistic geographers turned away from the scientific-positivistic *reduction of geography to geometry* – the conversion of 'real' geography into the tidy dots, lines, circles, hexagons and physical distances of the 'isotropic plain' – and instead insisted on thinking more philosophically about how in an existential sense the very 'humanness' of human subjects is bound up with the worldly spaces, places and environments they cannot help but occupy (see Entrikin's (1976) account of an emerging preoccupation with 'existential space'). There are several strands to the resulting *existential geography*, but the earliest manoeuvre of any substance was probably the attempt by Marwyn Samuels (1978, 1981) to found 'a spatial ontology of man' (*sic*) inspired by Buber's 1957 arguments about 'spatiality, distance and relation':

> Spatiality, [Buber] argued, is 'the first principle of human life' and entails a twofold process which he identified as (1) 'the primal setting at a distance' in order to (2) 'enter into relations'. . . . That initial ability, 'the primal setting at a distance', constitutes the ontological ground of any human existence. . . . As Buber argued, however, such detachment does not alone suffice to explain spatiality. Rather, detachment has a purpose, which is to say the second part of spatiality, 'an entering into relations'. Relationship is not conceived here as the opposite of distance, but rather as the goal of the 'primal setting' apart.
>
> (Samuels, 1981, p. 117)

The claim here is that human beings (unlike any other life form or being) have the intellectual capacity to distinguish in an abstract sense between themselves and other phenomena, and this is the process of 'detachment' whereby people conceive of themselves as 'distanced' from the world, but that in so doing they also evaluate more or less consciously the nature of their relationship to other phenomena by thinking in terms of how 'near' or how 'far' these phenomena are from their own personal 'centres'. People might conceive of slugs in their garden as being hugely different and 'distant' from themselves, for instance, whereas they would probably conceive of their best friends on holiday in another country as being very similar and 'close'. The space under discussion by Buber and Samuels is obviously not the physical space measured in kilometres or miles, then, and rather it is the sort of space replete with existential meaning that appears in Tuan's thoughts about geography and human nature (see above, and note that Tuan was explicit about the existential leanings of his 1971 paper).

There are additional features to Samuels' existential geography we shall return to presently, but what we must mention finally in this section is that a more pervasive existential foundation to the reconceptualisations undertaken by humanistic geographers has been the philosophy of Heidegger. For Heidegger, the central concern was that of Being (with a capital 'B') – the Being that 'lies in the fact that something is'; that it *is* 'in reality, in presence-at-hand, in substance, in validity' (Heidegger, 1962, p. 26) – and the key to unlock this mystery, so he argued, resides in an entity called *Dasein*, which is 'distinguished by the fact that, in its very Being, that Being is an issue for it' (p. 32): 'to Dasein, Being in a world is something that belongs essentially. Thus, Dasein's understanding of Being pertains with equal primordiality both to an understanding of something like a "world", and to the understanding of the Being of those entities which become accessible within the world' (*ibid.* p. 33). A passage such as this is usually taken to indicate that an entity called Dasein – an entity that turns out to be nothing more nor less than the *human being* itself – possesses, in the depths of its own sense of 'being-in-the-world', the conceptual materials through which it is possible to glimpse the universal truth of Being and also the more specific truths attached to the Being of particular non-Dasein (or non-human) entities. In outline it appears as if Heidegger was chiefly interested in *time* – and his greatest work is entitled *Being and Time* – but it should be clear from the above quotation that his entire philosophy launches from the question of human existence in a worldly environment full of things that, by virtue of their very disposition in *space*, are 'accessible' to both the 'use' and the 'concern' of human beings. Without wishing to become emboiled in the difficulties of Heidegger's arguments about 'spatiality' as pursued in the earlier sections of *Being and Time* (Heidegger, 1962, especially pp. 134–48; see also Relph, 1985; especially pp. 17–19; Pickles, 1985, Chapter 9; Soja, 1989, Chapter 5), what we can say here is that Heidegger supposed the Being of entities to be fundamentally bound up with both their actual distance away from human beings who might use them, and their perceived distance (as in Buber's existentialism) away from human beings who might have a concern for them. Understood in this light, Heidegger's philosophy challenges us to rethink the whole of human geography in a manner that sees space, place and environment as profoundly implicated in the 'biggest' philosophical questions about the very Being of both human beings and everything else 'accessible' to human thought and action. Humanistic geographers such as Buttimer, Pickles, Relph, Seamon and others evidently recognise and strive to meet this challenge, and Pickles (who is the most philosophically 'pure' in his turn to phenomenological and existential materials) concludes that geography – and here he means Relph's 'academic geography' – must follow Husserl's procedures and Heidegger's insights into the spatiality of existence in order that 'geography *can* be a *human* science of human spatiality' (Pickles, 1985, p. 170).

The elaboration of a geographical humanism

People and place
What should now be recognised is that the various exercises in phenomenological and existential geography discussed so far not only signposted a rethinking of basic geographical concepts, but also began to consider – and on occasion

to back up with more 'empirical' observations – the way in which 'ordinary' people leading 'ordinary' lives encounter, perceive and perhaps reflect upon the spaces, places and environments all around them. To be sure, the main purpose of these exercises has been to tease out the 'transcendental' (universal, timeless, placeless) essences supposedly embodied in how all people experience space, place and environment, in part as an end in itself and in part as an input to the reconceptualisation described above, but what has also surfaced on occasion is a genuine concern for the more 'everyday geographies' of the places in which we live and labour: for the houses, streets, factories, offices, schools, fields, parks, cinemas and so on where we spend most of our days, and about which we unavoidably develop a *sense of place* – a rudimentary understanding of how this place 'works' and a nagging feeling towards this place of liking, disliking, loving, hating, accepting, rejecting or whatever. And it is in this respect, so we would argue, that humanistic geography has probably made its most significant contribution to human geography: not in directing the attention of a few researchers to the deepest phenomenological and existential connections people have with their places, but in sensitising numerous researchers (many of whom would not even begin to label themselves as humanistic geographers) to the everyday and yet often quite intimate attachments all sorts of people (and not just philosophically inclined scholars) have to the places that encircle them. And it also in this respect that we want to talk about an emergent and innovative *geographical humanism* (see also Ley's (1981a) discussion of what he terms 'geographic humanism') that in various ways abandons the strait-jackets of pre-existing philosophical positions to make its own distinctive contribution.

There are a number of components to this geographical humanism, one of which involves the slightly off-beat engagement between 'humanistic geography and literature' (see Figure 3.5), but it is necessary first to underline the point that the research entailed here signals an important sea-change in humanistic geography from concentrating on the humanity of the researcher (and hence on epistemological questions about how human subjects can acquire knowledge of geographical objects) to focusing directly upon the humanity of the people the researcher studies. For somebody like Pickles this shift in emphasis is problematic, since he worries that the rigour of a phenomenological geography designed to 'ground' the discipline in a proper understanding of basic geographical concepts becomes swamped by an ill-defined *geographical phenomenology* 'concerned with lifeworld *as object* of study and as everyday mundane lived experience' (Pickles, 1985, p. 45). In other words, he urges an interest in 'lifeworlds' (see Figure 3.3) for the researcher seeking clues about the deeper essences of things geographical, and is thereby dubious about the tendency of other researchers who are prepared to anchor their inquiries very much in the 'here and now' of these lifeworlds as they are immediately constituted in the everyday experiences of ordinary people:

This [tendency] has fostered a widely-held misconception that a phenomenological geography takes the lifeworld as its object of study and that phenomenology has limited appeal only to certain branches of geography. This appeal is seen to be primarily in the area of historical, cultural and behavioural research, at the micro-behavioural level, not at the macro-spatial level. The emphasis is said to be on the cognitive states of the people involved, which should make the approach immediately attractive to social and cultural geography.

(*ibid.* p. 60)

Figure 3.5   Humanistic geography and literature

A specialised field of inquiry emerging as one consequence of the encounter between geography and the humanities is the interest a number of humanistic geographers are showing in imaginative litera-ture, and here it is argued that there are often important geographical insights to be gleaned from 'literary revelation' (which supposedly con-veys a 'truth' that goes 'beyond the mere facts' of non-fictional 'repor-ting'). An important collection of essays in this respect is Pocock's (1981a) *Humanistic Geography and Literature*, in which Pocock sug-gests (p. 12) that 'broadly, the geographer's engagement with literature in his [*sic*] study of places varies along a continuum between landscape depiction and human condition'. What Pocock means by this is that geographers are concerned both with imaginative 'word paintings', the ability of writers to weave pictures in words that somehow capture the 'flavour' of a place, and with the more profound sense that writers occasionally convey of how the humanity of a particular people is inti-mately bound up with the natural and built surroundings in which these people exist. Geographical studies of literature have now become quite commonplace (see, for instance, Mallory and Simpson-Housley, 1990), although it is important to note Daniels's (1985, p. 149) criticism of these studies for stressing the privileged and 'mysterious' abilities of the poet/novelist over and above the well-known 'literary conventions' that actually order most literary evocations of place. What might also be mentioned here is that some of the most recent (and arguably more 'mature') engagements between geography and the humanities have revolved not so much around literary senses of place as around the representation of landscapes – and notably of the 'social relations' codified in these representations – found in all manner of artistic pro-ductions, including both architecture and landscape design itself (see the essays in Cosgrove, 1982; Cosgrove and Daniels, 1988).

Our view is that the tendency towards making everyday lifeworlds into 'ob-jects of study' certainly does have an 'appeal' to 'certain branches of geogra-phy' such as social and cultural geography, and we will amplify this claim below, but where we differ from Pickles is in not seeing any great problem in what is effectively the sliding away of substantive inquiries from the strict ambitions and principles of his phenomenological geography. Rather, it seems to us entirely healthy that practising humanistic geographers should take their inquiries into waters that may not be philosophically pure, but that arguably begin to address the everyday attachments of people to places in a fashion that both resonates with the thinking of 'the people' themselves and illuminates the emotions lying behind so many contemporary struggles (many of them very political) organised around places as small as hamlets and as big as nations.

It also seems to us that most humanistic geographers would share our view that we gain more than we lose by developing a geographical humanism (which is basically the same as what Pickles calls a geographical phenomeno-logy), and we think that this is even the case for writers such as Relph and

Samuels. Indeed, Relph is perhaps best known for his path-breaking 1976 text, *Place and Placelessness*, which clearly combines an alertness to everyday people–place relations with a 'stricter' phenomenological search for geographical essences.

> My purpose in this book is to explore place as a phenomenon of the geography of the lived-world of our everyday experiences. I do not seek to describe particular places in detail, nor to develop theories or models or abstractions. . . . [M]y concern is with the various ways in which places manifest themselves in our experiences or consciousness of the lived-world, and with the distinctive and essential components of place and placelessness as they are expressed in landscape.
>
> (Relph, 1976, pp. 6–7)

In practice what Relph does is 'to demonstrate the range of place experiences and concepts' (*ibid.* no pagination), and in so doing to isolate in both himself and others a series of everyday reactions to place – and here he offers a sort of typology – which are identified with names such as 'rootedness', 'rootlessness', 'insideness' and 'outsideness'. Alternatively, whilst Samuels' initial ambition was to employ existentialism in rethinking the moorings of geography in notions of 'distance and relation' (see above), his later work widens out into what he terms 'an existential geography' – although the term *geographical existentialism* might be more in keeping with our own use of terminology here – concerned with the way in which meanings are both instilled into and distilled from the interweaving of a people's history with a landscape's 'biography' (and in this context 'landscape' and 'place' can be regarded as interchangeable terms):

> ultimately an existential geography is a study in the biography of landscape. The two most important aspects of that geography are (1) it begins with the subjective or with the issue of authorship in order (2) to discover the relations individuals and specific groups have with their environments as objects of concern. A biography of the former is here always a history of the latter. Effectively, an existential geography is a type of historical geography that endeavours to reconstruct a landscape in the eyes of its occupants, users, explorers and students in the light of historical situations that condition, modify or change relationships.
>
> (Samuels, 1981, p. 129; see also Samuels, 1979)

These arguments from Relph and Samuels are certainly attractive ones, in part because of the evocative language in which they are couched, but what we would suggest in addition is that the 'power' of their claims lies largely in the recasting of philosophical concerns to cope with the 'real' complexities of everyday attachments between 'real' peoples and 'real' places. (And this 'recasting' might be seen as an escape from the tyranny of foundational philosophies that is itself necessitated by a need to deal directly with the geography of peoples and places; see also Chapter 6 and the reasoning in Curry (1982a, 1982b)).

Practising and retheorising humanistic geography

It might be objected that the sorts of studies advocated and conducted by the likes of Relph and Seamon still retain an overlying philosophical flavour, or at least retain a tendency to write in a curiously 'detached' tone that imposes its own order – the 'readings' of its own authors – upon the details of particular peoples in particular places (and note Relph's (1976, p. 6) acknowledgement that he does not 'describe particular places in detail'). We would add the rider here that Relph, Samuels and others should not be criticised for not doing what they never set out to do, but what we will indicate is that it is only in the most obviously substantive of humanistic geography inquiries – those that really do strive to recover the 'nitty-gritty' of everyday people–place relations – that we find the seeds of a geographical humanism arguably central to the broader 'remaking' of human geography (see Kobayashi and Mackenzie, 1989, where this 'remaking' is carried forward through 'a dialogue *between* humanism and historical materialism in human geography'; see also our Chapter 4).

Now, it is important to recognise that one route taken by a few humanistic geographers has been towards the realm of highly *personal and subjective geographies*: a route taken by those who feel that properly humanistic research must try to get as close as possible to the place experiences of individuals as individuals, and who thereby use very intensive methods: 'encounter groups'; in-depth and repeated interviews; joint writing of personal biographies and scenarios, in order to tease out precisely how 'Jean', 'Sam', 'Nelson' and other nameable individuals interact with their encompassing places. The work of Graham Rowles (1978a, 1978b) is revealing in this connection, given that in his attempt to move beyond the 'navel-gazing' of the existing humanistic geography literature he uses 'experiential field work' – a methodology predicated upon establishing close if often awkward 'personal relationships' – in exploring the geographical experiences of five elderly people living in Winchester Street, an inner-city neighbourhood of an eastern US city. And more recently, Paul Rodaway (1988) has used the methodology of 'group reflection' in bringing together a small number of locals living in a Co. Durham village to 'open' their 'environmental experience', a process that is supposed to occur over a long period of time as the individuals discuss with the researcher and among themselves (these discussions being taped for the researcher to analyse at a later date) how they perceive the significance of their own 'experiential geographies'. Other examples of this very definite 'personalising' of humanistic geography could be mentioned, of course, and it is interesting that in looking for theoretical and practical guidelines for the inquiries being conducted researchers have gradually moved away from philosophical debate towards more psychological and psychoanalytical materials (see especially Burgess, 1988).

Whilst Rodaway, Rowles and others suggest that their approach does allow some 'generalisation' whereby themes common to the geographical experiences of different individuals are extracted, their primary interest in individuals as individuals does not match up all that well to the more usual interest of human geographers in the 'macro-scale' patterns of many people living in sizeable places. As Peter Gould (1976, p. 87) once asked rather sarcastically,

Why, as geographers, are we interested in such individualistic details? And how much further do we push back – to psychoanalytic studies that tell us that some 'Lolits' (Little Old Ladies in Tennis Shoes) avoid certain stores because they have red shutters, and that as children they had a traumatic experience in a house with red windows?

Gould's question was prompted chiefly by certain trends within behavioural geography (see above) that were then focusing increasingly on individuals rather than on aggregates, but he clearly – if a little nastily – anticipated a difficulty that is now present in those humanistic-geographical studies where researchers are indeed 'pushing back' into the domain of psychoanalysis. However, this route is definitely not the only one that has been taken by humanistic geographers striving to recover the 'nitty-gritty' of everyday people–place relations, and we want to conclude this chapter by tracing the emergence of a geographical humanism that takes seriously the irreducibly *inter-personal and inter-subjective geographies* of human groupings whose shared 'world views' – whose common values, meanings or ideologies – are intimately bound up with the material places of which they are quite literally 'a part'. Accusations of 'whimsicality', of 'romanticism' or of overstating the extent to which human agents are free to shape their own destinies (all of which have been levelled by Marxist and 'agency-structure' theorists; see Chapters 2 and 4) may have some validity when applied to other variants of humanistic geography, but we believe that they lose much of their force when directed at inquiries where the root *raison d'être* is the recovery of thoughts and actions shared by human groupings living, working, defending and re-building the places of their worlds. In order to give some flavour of what is involved in these inquiries we will first provide a brief account of David Ley's work on the social geography of the city (see also his own textbook on this social geography: Ley, 1983).

There can be little doubt that Ley has been one of the foremost figures in humanistic-geographical endeavour over the last twenty or so years, particularly given that he co-edited the *Humanistic Geography: Prospects and Problems* volume (Ley and Samuels, 1978a) and has written a number of important theoretical pieces calling for the more sensitive incorporation of human agency into geographical research (Ley, 1980a, 1981a, 1981b, 1982). But he also recognises that 'too often the purity of philosophical discourse has run aground upon the rocks of geographic practice' (Ley, 1988, p. 122), and as a result he clearly seeks to build away from strict foundational philosophies such as phenomenology and existentialism – and note his remark that Husserl's phenomenology 'is now little more than a curiosity' (Ley, 1981a, p. 224) – towards 'humanistically-orientated work in social geography' (*ibid.* p. 217; see also Ley, 1977, 1978). And the latter is clearly concerned not with transcendental essences nor with other forms of scholarly abstraction, but with the everyday and usually 'taken-for-granted' meanings of social actions available to social groups in specific social contexts. In practice as well as in theory these concerns have repeatedly taken Ley back into the maelstrom of city people and places, and a significant marker here both for him (see the reflections in Ley, 1988) and for humanistic geography more generally is the early research he undertook on 'The black inner city as frontier outpost' (1974). In this remarkable study he thought of himself as an 'explorer' entering the 'wild' lands of inner-city Philadelphia (and he both played with and criticised these metaphors), and there – confronted by the real difficulties of 'getting by' in this space so alien to middle-class white observers – he exposed the errors and the prejudices associated with conventional depictions of people and place in black urban America. Moreover, he tried to offer a more sensitive account of what was actually 'taking place' in the neighbourhoods he studied, and (to draw upon but one aspect of his findings) he examined the intense attachments street gangs developed for their own 'gang spaces', 'turfs' or 'territories' as symbolised in the use of such 'territorial markers' as 'graffiti obscenities':

graffiti markings represent the language of space for members of the street gang culture. Where territories meet, space is most highly contested, and aggressive behavior is most appropriate. With increasing proximity to the core of the turf, the meaning of space changes, and there is an orderly decrease in assertive behavior, until at the core security is perceived to be maximal. In this zone, where threat is regarded by gang members as unlikely, assertive behavior against rivals becomes unnecessary, and graffiti obscenities are almost absent.

(ibid. pp. 218–19)

Figure 3.6   Pragmatism and geography

Pragmatism is a philosophy that arose around the turn of the century as a 'movement that aimed to ground philosophical activity in the practicalities of daily life' (S. J. Smith, 1984), and as such it sought to negotiate the thorny and seemingly irresolvable 'big' questions of philosophy by examining instead both the constitution and the application of 'knowledge' in everyday life. The intellectual roots of this philosophy lie in the writings of North Americans such as Dewey, James and Peirce, although more recent claims for pragmatism can be found in Rorty's insistence that knowledge should be understood not as offering some unobtainable 'accurate representation of reality' but as nothing more nor less than an ongoing 'conversation' between *all* of us – and not just between scholarly philosophers – with our different things to say moored in different circumstances for saying them (see Rorty, 1980, 1982). Pragmatism was first discussed in the geographical literature by Frazier (1981), who outlined how 'applied geographers' – perhaps geographers involved in housing schemes – should draw 'pragmatically' upon a range of scientific and humanistic theories, models and techniques in the course of fulfilling their specified 'action roles'. But at the same time he (*ibid.* p. 68) noted that 'human interests, desires, prejudices and group values vary across space', and in so doing he took seriously the influence of geographical variations in everyday contexts for both knowledge-creation and knowledge-implementation. And this sensitivity is one that surfaces in the better-known geographical arguments for pragmatism associated with Jackson and Smith (1984, Chapter 4; S. J. Smith, 1984), who are attracted to this philosophy because it informed the studies of the early twentieth-century Chicago urban sociologists – notably the 'ethnographic' studies through which Park, Burgess and others explored the lives and self-understandings of ordinary Chicagoans – that are currently stimulating the renaissance of social and cultural geography (see later). The argument here has several strands, but on the one hand it suggests that human geography should foresake 'grand' claims to knowledge because the knowledge with which it deals and which it can meaningfully obtain in practice is itself geographically fragmented (see Chapter 6), and on the other hand it suggests that in the absence of 'grand' philosophical guidelines for practice the discipline must (re)gain a sense of moral and political purpose to justify what its practitioners do in the fields and streets of the world.

Ley concluded that here could be seen very graphically 'the importance of an existential space in determining behavior' (*ibid*. p. 219), but what should immediately be added that in no way was the study of gang space a self-conscious exercise in some 'pure' existential geography (and note that its arguments could equally well be paired off with more ecological notions of 'territoriality' or with more sociological notions about conflict and power). The simple point we are underlining is that Ley's 1974 research was excited principally by very real, complex and sometimes tragic substantive issues written into the people–place relations of black urban America; and as such, Ley strove to interpret these relations as best he could, using whatever theories (including much 'Chicago sociology'), methods (chiefly 'participant observation') and sources as seemed appropriate to the task in hand (see Figure 3.6 on the philosophy of *pragmatism*).

In many senses Ley's humanistic geography does indeed 'squib out' from underneath the stricter templates of those philosophies that are often regarded as the theoretical baseline for humanistic-geographical practice, but what must now be acknowledged is that in various places Ley (1977, 1981a; see also Gregory, 1978, pp. 133–44) attempts to ground his substantive interests in the *constitutive phenomenology* of Alfred Schutz (see Figure 3.3): a theoretical position Schutz developed as a conscious attempt to preserve the spirit of Husserl's philosophy, but in which he abandoned Husserl's obsession with buried essences so as to render phenomenology more compatible with the concerns of social scientists. And it was through Schutz's intervention that phenomenology became recast as the study of everyday and 'taken-for-granted' lifeworlds *per se* rather than as the probing of lifeworlds for clues regarding deeper essences (precisely the manoeuvre that Pickles objects to; see above), and this is how Ley (1977, pp. 504–5) begins to present what is involved here to geographers:

> As social geography follows its agenda and dips beneath spatial facts and the unambiguous objectivity of the map, it encounters the same group-centred world of events, relations and places infused with meaning and often ambiguity. Husserl, in his later writing, characterised this realm as the *lifeworld*. More recent philosophers like Schutz and Merleau-Ponty have urged that this reality encompassing mundane experience is not irrational and impossible to study.

But perhaps the most attractive feature of Schutz's stress upon 'mundane experience' for a *social* geographer such as Ley is its insistence on the sociality of everyday life: its insistence on taking seriously the inter-subjectively constituted lifeworlds – the shared meanings and 'common-sense knowledges' – associated with groups of people who lead similar lives under similar circumstances in similar places. Whilst Ley appreciates that places as physical assemblages of homes, workplaces and roads may contain social groups whose lives are actually very different, and may thereby present themselves as arenas of 'multiple realities', he still supposes that groups localised in particular places will often constitute similar lifeworlds and that a crucial ingredient of these lifeworlds will almost always be an identification with their 'host' places. Place, group and lifeworld are hence seen as three closely linked entities that should stand at the heart of a humanistically-orientated social geography, and when Ley urges us to draw upon Schutz's phenomenology to furnish us with these ingredients for substantive geographical inquiry he neatly complements the call to use Schutz in reformulating sociological inquiry (see Berger and Luckmann, 1967).

As already indicated, we think that Ley's version of humanistic geography negotiates many of the criticisms levelled at other studies in which philosophical or even 'élite' cultural concerns (as in the turn to literature; see Figure 3.5) have swamped a concern for more 'sociological' matters, but brief mention should still be made here of the 'critical edge' that remains between humanistic geography – even when conducted by Ley – and Marxist geography. Here, for instance, is Richard Walker and Douglas Greenberg (1982, p. 29) discussing the flaws they detect in Ley's 'eloquent espousal of humanistic principles in geography' as a counter to positivism:

> Nevertheless, he has not broken with the empiricist ontology that lies behind positivism. He does not distinguish between events, everyday life, individual circumstance, and their causes, which are normally unobservable and unaltered by the contingent situations of everyday circumstance. . . . Similarly, he takes the ideological pronouncements of human subjects, such as the 'anti-business' rhetoric of the urban reformers in Vancouver, at face value, instead of catching the refraction of consciousness off the hard surface of material reality.

In other words, Walker and Greenberg suppose that Ley's humanistic geography still gives too much privilege to human beings – to the everyday events, experiences and interpretations of people leading everyday lives – and in so doing remains 'empirical' and 'superficial': and for them, alternatively, the satisfactory way to proceed is one that views people's existence and self-understanding as fundamentally shaped by 'the hard surface of material reality'. Moreover, Walker and Greenberg suppose that this reality is itself underpinned by deeper 'causes' and 'logics', and in this respect they subscribe to the philosophical tenets of *realism* (see Chapter 5), and their own inclination is to use Marxist concepts in which causes and logics are reckoned to be sedimented in the invisible and largely invariable workings of the (capitalist) mode of production (see Chapter 2). A valuable warning is certainly posted here about the deficiencies of a humanistic geography that closes itself solely around the thoughts and actions of 'ordinary folk', and that thereby ignores the very real constraints (particularly economic and political ones) that limit the extent to which many peoples in many places can control the direction of their own lives. But it should also be added that humanistic and Marxist geographers are usually asking very different questions about social life, the former tending to inquire about social meaning and the latter tending to inquire about social inequality, and as such we would not want to use the sorts of arguments pursued by Walker and Greenberg to condemn everything attempted by humanistic geographers (just as we do not think that the humanistic arguments pursued by Duncan and Ley (1982) can condemn everything attempted by Marxist geographers). Rather, we feel that both humanistic geography and Marxist geography open important windows on the human world, and that on some occasions it will be appropriate to employ either humanistic or Marxist concepts to guide substantive inquiries whilst on other occasions (and especially when theorising in *structurationist* terms; see Chapter 4) it will be appropriate to find ways of combining both humanist and Marxist concepts.

Meaning, knowledge, language and local geographies
We have spent some time discussing Ley's work – indicating something of both his substantive concerns and his theoretical deliberations – because in various

ways it can be taken as emblematic of an important development sweeping throughout human geography and its thematic subdisciplines. And in saying this we do not simply mean that Ley's sensitivity to people in their places is now being duplicated in countless geographical studies, although this is undoubtedly the case (and most notably in a text such as Western, 1981); rather, we mean that Ley is effectively pursuing arguments about meaning, knowledge and even language as (in many cases) very much *local affairs* – as local affairs through which particular people in particular places gain a sense of identity and purpose that, even if little reflected upon and not entirely 'positive', enters centrally into their practices and politics – and in so doing is clearly paralleling a host of other substantive and theoretical interests currently exciting human geographers. This is not to suggest that these other geographers emulate Ley in grounding their ideas in Schutz's phenomenology, but it is to suggest that the conceptual materials they do draw upon – and the intellectual terrain here is becoming increasingly complex – are at bottom compatible with Schutz's theoretical insights (or, better still, are compatible with the use Ley makes of Schutz's insights in his humanistically-orientated social geography).

Ley often refers to his work as 'social geography', and it is perhaps in the subdiscipline of social geography (which has traditionally studied social groups in villages, towns and cities) that a geographical humanism concerned with people–place relations has proved most influential. It is therefore instructive to read Peter Jackson and Susan Smith's influential 1984 text, *Exploring Social Geography*, in which they engage with the literature of humanistic geography when outlining what they term a 'hermeneutic revival' in social geography (*hermeneutics* is the interpretation of meaning; see Figure 3.7): 'a hermeneutic revival in which social geography re-emerges as an exploration of the relationship between people and place, seeking to go beyond empathetic understanding to offer an *interpretation* of human experience in its social and spatial setting' (Jackson and Smith, 1984, p. 20). Jackson and Smith are sympathetic towards the elements of critique and reconceptualisation present in humanistic geography (to use our own terminology), but they insist, as does Ley, in pressing beyond both the strait-jacket of philosophy and the focus on individuals to argue for a social geography alert to the ways in which *communities* of people *intersubjectively* build up an understanding of how their local worlds 'work' (an understanding that is largely implicit but that is in a sense 'represented' in their practices). And at this point Jackson and Smith echo what Ley (1981b) had speculated about three years earlier, in that they begin to think of these intersubjective understandings as dimensions of the *culture* sustained by particular people in particular places. This means that they connect up social geography to cultural geography, but now a cultural geography concerned less with material artefacts and technologies – the traditional interest of North American cultural geographers such as Fred B. Kniffen and Carl Sauer – and more with immaterial modes of thinking and living (see the accounts given in Cosgrove and Jackson, 1987; Cosgrove, 1989a; Jackson, 1989).

Moreover, a significant input to the resulting fusion of social and cultural geography becomes the 'interpretative anthropology' of Clifford Geertz (1973, 1983), whose approach to culture is basically 'semiotic' in regarding it 'as a series of signs and symbols which convey meaning' (Jackson and Smith, 1984, p. 38; see also Richardson, 1981, S. J. Smith, 1988). There is a double sense in

Figure 3.7 Hermeneutics

A broad definition of hermeneutics sees it as an intellectual tradition that 'emphasis(es) the symbolic constitution of the social-historical world, the way in which this world is created by speaking and acting individuals whose creations can be understood by others who partake of this world' (Thompson, 1984, p. 10). It is a tradition that originated in the examination of *meanings* contained in biblical and other theological texts, but its brief was soon extended to the study of all manner of 'texts', including that 'text' that is nothing but the everyday events and practices of 'social life' itself. This manoeuvre notwithstanding, though, what is distinctive about hermeneutics – as opposed to, say, phenomenology (see Mugerauer, 1985) – is its concern for language *per se* (for words, dialects and languages) and for regarding even non-linguistic phenomena as media that can communicate meaning to human interpreters.

which this is the case, in that the 'signs and symbols' – the things said, the things done, the gestures, the artistic products and so on – of a particular culture 'convey meaning' both to the people sustaining that culture and to the outsider (the researcher) seeking to interpret that culture. In the first instance, then, Geertz imagines culture acting much like Schutz's lifeworld: as the 'glue' of locally established meaning and knowledge through which particular people in particular places 'make' and 'remake' their lives, but the crucial inflection that Geertzian anthropology introduces is the notion that these meanings and knowledges are themselves born upon, are in a sense indistinguishable from, the manner in which they are spoken, 'talked', written, drawn, danced or otherwise communicated. What is signposted here is the need to pay attention to the detailed geography of intersubjective communication, and thereby to discover the myriad connections running between 'what things are said', 'how things are said' and the people- and place-specific contexts in which they get said. Putting matters like this immediately suggests links with projects couched in slightly different vocabularies, and here we have in mind both Michael Curry's borrowing from Wittgenstein to propose studying the 'ordinary languages' bound up with specific 'forms of life' (see Curry, 1982a, 1982b; see also Barnes and Curry, 1983) and Robert Mugerauer's call for a 'hermeneutical' approach to the regional geography of words, dialects and languages (and to regional variations in 'the saying of the environment'; see Mugerauer, 1981, 1984, 1985):

> Words, including names, are not merely labels, but the evocation of what things are and [of] how they are related to other things in the web of particular lives and places. And more than names and words, there is langauge itself, which is not any fanciful artifact, but that which has the power to articulate and join humans to plants, animals and activities in a surrounding world. The entire fabric of a people's meaningful world – the total environment – comes along with the whole of that people's language.
>
> (Mugerauer, 1985, p. 59)

At the same time this nexus of arguments dovetails intriguingly with the insistence on taking seriously what Allan Pred calls (1989a) 'the language of everyday life', which he reckons to possess a definite geography anchored in the differing economic, social and political circumstances of different places:

Words spoken in place,
Meanings made in place,
Words and meanings here, but not there.

(Pred, 1989b, p. 211)

What should now be apparent is that there is an exciting if somewhat bewildering jumble of ideas involved here – a jumble of ideas about local meanings, knowledges, languages, peoples and places that is inspired by (although is certainly not restricted to) the input from Geertzian anthroplogy – and it is obvious that this 'heady brew' takes us some way from the realms commonly identified with humanistic geography.

A second aspect of how culture 'conveys meaning' revolves around how outsiders, and more particularly researchers, 'interpret' the 'signs and symbols' with which they are confronted, and this is why Geertzian anthropologists see their task as one primarily of 'reading' and then 'writing' culture. Geertz himself depicts anthropology as a complex intellectual activity involving the researcher in time-consuming and often heart-breaking *ethnographic* inquiry 'in the field' – the many hours spent among 'native' peoples observing them, interviewing them, participating in their lives and rituals – and then involving attempts to understand and to represent the worlds of these peoples in terms as faithful as possible to their own interpretations: 'Doing ethnography is like trying to read (in the sense of "construct a reading of") a manuscript – foreign, faded, full of ellipses, incoherencies, suspicious emendations and tendentious commentaries, but written not in conventionalized graphs of sound but in transient examples of shaped behavior' (Geertz, 1973, p. 10). The appeal of 'doing ethnography' is one that is currently interesting numerous social and cultural geographers, many of whom are looking for a new kitbag of 'qualitative methods' with which to supplement the more 'quantitative procedures' favoured by spatial science (see Eyles and Smith, 1988), and what is also notable in this connection is the recovery of that alternative tradition of the so-called Chicago 'school' of urban sociology – the 'school' that produced the concentric-ring models of urban land use so popular in student textbooks – which saw numerous researchers going out on the streets of Chicago to conduct 'urban ethnographies' of the people (the blacks, the Sicilians, the hobos, the delinquents and the prostitutes) they found there (see S. J. Smith, 1981; Jackson and Smith, 1984, Chapter 4; Jackson, 1985). Chicago ethnographies such as Gerald Suttles's *The Social Order of the Slum* (1968) were an important influence upon Ley's Philadelphia study, for instance, and what is more the 'symbolic interactionism' of these ethnographies – the 'assumption that "reality" is a social production, consisting of social objects whose meanings arise from the behaviours that people direct towards them' (Jackson and Smith, 1984, p. 81) – have clearly informed the thinking of both Ley (who points to connections between this position and the arguments of Schutz) and other social-cultural geographers (notably Duncan, 1978; Eyles, 1989). Thanks to the arguments of Geertz, though, these geographers now fully appreciate the difficulties associated not just with the practicalities of ethnographies but also with the 'mediation' involved as a researcher brings his or her 'frame of reference' into contact with those of the people being researched (as the humanity of the researcher mingles with the humanity of the researched). It is now widely recognised that the researcher cannot ever hope to gain access to the *truth* of other people's lives, and therefore cannot expect to do more than describe these lives as carefully and in as much detail as possible (a process Geertz calls 'thick description') and in so doing acknowledge that the descriptions being produced are really nothing but 'stories' or 'fictions' ('something fashioned', the original meaning of *fictio*). But this is not to denigrate what can be achieved, and neither is it to imply that 'doing ethnography' is without rigour or academic merit: rather, it is to engage very honestly with both the enchantment and the problems associated with researchers trying to gain an insight into the worlds of other peoples in other places, and it is also to insist that this insight emerges not by supposing (as might a philosophically inclined phenomenologist or existentialist) that these peoples are basically the same as us but by letting them and their version of humanity simply be different (see also Chapter 6). And the purpose Geertz ascribes to interpretative anthropology is surely one that can and should be taken on board by practitioners of an 'interpretative human geography' (Ley, 1988; D. M. Smith, 1988): 'The essential vocation of interpret[at]ive anthropology is not to answer our deepest questions, but to make available to us answers that others, guarding other sheep in other valleys, have given, and thus to include them in the consultable record of what [humanity] has said' (Geertz, 1973, p. 30). This purpose for human geography is admittedly far removed from the original stated intentions of humanistic geography, but we suspect that most of the humanistic geographers mentioned above would have considerable sympathy for geographical inquiries into the worlds of all sorts of peoples in all sorts of places (whether Aborigines in Australia, hill farmers in Wales, academic geographers in Lampeter, or whatever) and would see value in such inquiries running alongside – and maybe contributing to or perhaps learning from – their own more phenomenological and existentialist projects.

4

STRUCTURATION THEORY: ANTHONY GIDDENS AND THE BRINGING TOGETHER OF STRUCTURE AND AGENCY

A middle ground between structure and agency?

> Several fashionable movements come quickly to mind: Giddens and structurationism, Hägerstrand and time-geography, Foucault and post-structuralism, Roemer and the rational choice Marxism, Kristeva and deconstruction, Bhaskar and realism, Habermas and communicative interactionism, Jameson and post-modernism, Althusser and structuralism, Sabel and flexible production. . . . Intellectuals are creatures of fashion, and this, too, affects the fate of ideas. The ebb and flow of fashion is, at least, exhausting and, at worst, quite pernicious to progess in social science.
>
> (Walker, 1989, p. 156)

> The difficulty in constructing a workable theory of action is to avoid on the one hand the determinism of the structural view, and on the other the idealism and hyper-individualism of some non-structural approaches.
>
> (Duncan, 1985, p. 178)

In Chapters 2 and 3 we have seen the development of Marxist and humanist approaches in human geography from a relative lack of sophistication and a relatively high level of dogmatic orientation towards 'structure' and 'agency' respectively in the early 1970s, through to the much more sophisticated position of the 1980s where the interconnectivity between 'structure' and 'agency' has been more fully explored within both the Marxist and humanist camps. In Chapters 4 and 5 we turn to the approaches of *structuration* and *realism* that, in effect, represent alternative conceptual terrains in which the structure–agency debates have been played out (both by human geographers in particular and by social theorists more generally).

To understand how structuration was conceived, and how human geographers began to engage with it, it is important to recognise that the original proposals for a 'structuration theory' by Anthony Giddens were founded on a critique of the *structural* and *interpretative sociologies* that were connected to, but not synonymous with, the Marxist and humanist traditions that surfaced in the human geography of the early 1970s. Given the increasing sophistication of these traditions through the 1970s and 1980s, a straightforward appropriation of Giddens's initial treatment of the 'two sociologies' (see Thrift, 1983) to critique more *current* developments within both Marxist and humanist geography may now run the risk of caricaturing rather than accurately capturing the achievements of (and also the differences internal to) these two sets of approaches. This being said, though, Giddens's opposition to any single-minded attachment to overly structure-orientated or overly agency-orientated accounts of social life clearly was appropriate to human geography in the early 1970s, and does still comprise a vital warning to human geographers working in the 1990s.

It is this background that makes the attitudes of Richard Walker (quoted above) and others an interesting platform for understanding how structuration has been received by human geographers over the last decade or so. It has tended to be those who are heavily committed to either Marxist or humanist themes, or indeed to the positivist (spatial-scientific) tradition that preceded them, who have most bemoaned those 'fashions and fads' of post-positivist human geography Walker lists in a rather derogatory manner. His conclusion that these newer 'isms' are pernicious to the progress of social science reflects a theoretical community under attack from those who seek to present a different discourse by combining aspects of previous discourses (thus adulterating the unadulterated and defiling the pure). It would not be an overstatement for us to suggest in this context that some Marxist geographers have regarded structuration theory as a betrayal of fundamental structural principles, whilst many humanistic geographers similarly feel that the complexity of human beings is obliterated by the structurationist tendency to regard them as 'agents' rather than as 'people'. There can hence be no doubt that these tensions were real issues for the geographers concerned and were fought over accordingly.

On the other hand it is what has been seen by some as the inflexibility of some parts of traditional Marxism and humanism which has led to the search for a 'middle ground' or 'alternative ground' on which to base the theoretical foundations for social science. As James Duncan (quoted above) indicates, theoretical purity can camouflage either *determinism* or *idealism*. Through the Marxist lens, the camera can be construed as only seeing social, political and economic constraints, and ignoring the purposefulness and spirit of individuals; through the humanistic lens the camera can be seen to catch the ideals and cultures of individuals and groups, but not the political and economic boundaries within which individual and group behaviour is put into practice. Somehow, if only the insights concerning both *structures* and *human agency* could be interlinked, then social science would have a more realistic grounding for its thinking and practising.

This chapter, on structuration, and the next, on realism, explore two related groups of theories that have moved into this middle ground between studying structures and studying human agency. The fact that some contemporary Marxists and humanists are reformulating their critical frameworks in something of a similar direction adds significance to the movement. Moreover, the working out of this middle ground has had important implications not only in terms of discussing how structures and human agency interact but also in a refreshing attempt to define more closely the varying natures of structure and agency in the particular circumstances researched by practising social scientists. This in turn, and in a fashion that has both intrigued and been refined by human geographers, has led to important progress in understanding how action in society is crucially 'set' in time and space. But, and to underline the above comments, the search for middle ground will be the subject of widely different interpretations by different geographers. Some will see it as a poverty-stricken compromise that loses the conceptual sensitivity of either Marxism or humanism by trying in some sense to bring them together (combining the uncombinable?). Others will argue that a combination of structure and agency in the structurationist format represents a conceptual advance 'bigger' in its importance than the sum of the component parts. In this chapter we present an account of the theory and practice of structuration that, it is hoped, will enable a further exploration of these different interpretations.

A structurationist school of thought?

It is perhaps useful here to reiterate the common suggestion that any theoretical position rests ultimately upon two significant philosophical components:

1. *Ontology*: 'the theory of existence or, more narrowly, of what really exists, as opposed to that which appears to exist but does not' (Quinton, 1988, p. 605).
2. *Epistemology*: 'the philosophical theory of knowledge, which seeks to define it, distinguish its principal varieties, identify its sources and establish its limits' (*ibid*. p. 279).

In seeking to break down the supposed polarities of Marxism and humanism, the two principal middle ground bodies of theory, structuration and realism, have, as it were, started out tackling different philosophical components. Realism (see Chapter 5) largely tackles the epistemological polarisation of previous approaches, while structuration has been largely ontological in orientation. Anthony Giddens, a leading theorist of structuration, hence explains his task thus:

> What I'm trying to do is to work on essentially what I describe as an *ontology* of human society, that is concentrating on issues of how to theorise human agency, what the implications of that theorising are for analysing social institutions, and then what the relationship is between those two concepts elaborated in conjunction with one another.
>
> (In Gregory, 1984, p. 124)

It should be stressed at the outset that structuration does not represent any kind of 'flash-in-the-pan' response to the perceived ontological deficiencies of Marxism and humanism. Indeed, elements of structurationist thinking can be traced back at least as far as the writings of Berger and Luckmann in 1967, and there has been a long and multi-faceted history of theoretical writing in this vein since then, including the work of Bourdieu (1977), Touraine (1977), Bhaskar (1979) and Dawe (1970, 1979). Along with Giddens, these writings represent what Thrift (1983) calls 'the structurationist school' and what Pred (1981) recognised as an emerging consensus in social theorising. It is also worth emphasising at this initial stage that structuration need not be viewed as a form of all-embracing and cohesive grand theory. Rather, as Soja (1989) describes, it started as a series of abstract propositions that have interacted with more empirical applications such that the original theorums have been reworked in a series of stages. Thus, structuration theorising is seen as 'spiral- or helix-shaped': the core theorums have led to investigations of particular aspects of modern societies and these have in turn informed a reformulation of the abstract propositions. Inevitably, then, the delving of different theorists into this helix has resulted in some internal diversity of ideas and applications.

For the purposes of this chapter, however, we will examine structuration chiefly through the texts of its best-known theory-monger, Anthony Giddens, whose account began in 1971 and continues today. Giddens is important to human geography not only as the key theorist of structuration but also because he has engaged with many geographical ideas (particularly relating to *time-geography*) in his continuing discourse. The scope of Giddens's work is excellently captured by Gregory (1989b, p. 185):

> it makes most sense to treat Giddens's writings as a *research programme* developed through a continuous dialogue between the theoretical and the empirical. . . . Giddens's project is not linear: one proposition does not succeed another in a unidimensional unidirectional sequence. . . . Structuration theory then appears as a loose-knit web of propositions, some more central than others, some spun more tightly than others. In contradistinction to networks in the natural sciences, structuration theory is clearly not directed towards the study of 'laws' . . . [I]ts development has been determined . . . by both coherence rules (relating to the structure of the network) and correspondence rules (relating to empirical observations). These have required a constant reworking of its conceptual fabric, so that structuration theory has, of necessity, been developed unevenly.

Gregory's analysis is quoted at length because it provides an essential context to Giddens's own assertion that he wished to fire 'conceptual salvos' into social reality rather than conjuring up one 'big theoretical bang' (see Bleicher and Featherstone, 1982).

Giddens's structurationist project: phase I

Anthony Giddens is Professor of Sociology and Fellow of King's College at the University of Cambridge. His prodigious output of texts on social theory (see Figure 4.1) reflects Gregory's idea of a research programme (see above) in which a continuous dialogue between the theoretical and the empirical produces and reproduces an unevenly developed network of concepts and propositions, rather than any 'grand slam' attempt to introduce a total theoretical treatise at one go. Giddens himself has become the focus of considerable attention, attracting both adulation and criticism, as with all charismatic intellectual figures. He is one of the few contemporary social theorists to be the subject of a biographical critique in his own lifetime (see Cohen, 1989; Held and Thompson, 1989; Clark, Modgil and Modgil, 1990) and he has stood at the centre of many published debates on social theory (see, for example, the symposium in *Theory, Culture and Society*; Ashley, 1982; Bleicher and Featherstone, 1982; Gross, 1982; Hirst, 1982; D. Smith, 1982; and Giddens's reply, 1982).

Giddens's own journey from the inital theoretical roots of structuration, to the more refined and specific propositions that have been made useful to human geographers through the translations of Gregory, Thrift and Pred (see below), has been both incremental and interactive. To understand the subtleties of structuration it is necessary to retrace this journey back to its original stages (Dickie-Clark, 1984), and to follow it through at least two major phases of theoretical construction, first up to 1980, and then beyond (Gregory, 1986).

Figure 4.1   The major works of Anthony Giddens

1971	*Capitalism and Modern Social Theory*
1976	*New Rules of Sociological Method*
1977	*Studies in Social and Political Theory*
1979	*Central Problems in Social Theory*
1981	*A Contemporary Critique of Historical Materialism, Volume 1: Power, Property and the State*
1984	*The Constitution of Society: Outline of the Theory of Structuration*
1985	*A Contemporary Critique of Historical Materialism, Volume 2: The Nation-State and Violence*
1989	*A Contemporary Critique of Historical Materialism, Volume 3: Between Capitalism and Socialism*

Giddens's early writings (1971) were principally concerned with criticising the social theories of Marx, Weber and Durkheim, and here his inherent dissatisfaction with the existing polarities of theory becomes clear. His 1976 book criticised the *interpretative sociologies* of those whose theoretical focus was solely upon human agency, and here he criticises the idea that society can be theorised as the product of unconstrained human action: that is, the *voluntarist* position. In his 1977 and 1979 books he turns his attention to what he sees as the over-*deterministic* fascination that *structural sociologies* – notably Marxist ones – have had with social structures and social systems. He is particularly critical of the idea that structure occupies a position of primacy over agency, and that structures act as boundaries to, or constraints upon, action. Thus, the starting point for Giddens is a deliberate, knowing, debunking of interpretative and structural sociologies, partly for their theoretical deficiencies, and partly because the institutional momentum in each of these theoretical groupings raised problems of the separate systems or empires of knowledge. Indeed, Giddens (1977, p. 2) admits that 'One of my principal ambitions in the formulation of structuration theory is to put an end to each of these empire-building endeavours'. Structuation theory, though, could not be constructed simply on the motive of criticising the seeming exclusivity of some versions of Marxism and humanism. What was sought too was a reconstruction of the importance of both structure and agency. So far as the latter is concerned, Giddens looked to incorporate into his social theory the notion of human action as something rationalised and ordered by people in the world ('human agents'). He thereby wanted to treat people as knowledgeable and capable subjects, not as the cultural dupes of structural determinism, and he wanted to construe the changing circumstances of social life as a 'skilled accomplishment' by these subjects. To this extent he considered that the propositions of humanistic thought had to be retained in any new social theory.

On the other hand, however, Giddens argues that his social theory needs to recognise and retain particular features of the reality of social life as he sees it – power, struggle, institutional organisation and the like – which have previously been emphasised by the structural sociology theses. Importantly, however, he stresses that structure is not to be conceived of simply as a constraining barrier to action, but rather as an enabling *involvement* in that action. In this way Giddens confirms that certain structural properties of social systems can be both a medium for and an outcome of social practices.

At this point, then, Giddens reaches the idea of the *dualities* associated with structure and agency, which we would identify as one, if not the, key characteristic of the structurationist school. He argues that the only ways in which to bring together these facets of structure and agency in a non-functionalist manner are

1. to recognise the *duality of structure*: that is, the manner in which structures enable behaviour, but behaviour can potentially influence and reconstitute structure; and
2. to recognise the *duality of structure and agency*: that is, to transcend the dualism of deterministic views of structure and voluntaristic views of agency.

As we have already seen in Chapters 2 and 3, there is nothing particularly new about this, but the notion of duality represents a key stage in Giddens's structuration journey. Nicky Gregson (1986, p. 185) explains the importance of duality as follows:

> Duality is central to the entire structurationist programme, figuring in both Giddens's presentation of the agency-structure relationship and the links between this and time and space. . . . [T]he interdependence between history, society and purposeful individual action is equal in weight; neither society, nor individuals are assumed to exert a greater influence on events than the other. The relationship between agency and structure in time and space is treated similarly; whilst temporal and spatial organisation limit individual action, they are, at the same time, the creations of history, society and individual action. Again, each exerts a determining influence on the other but this is again of equal weight.

Gregson goes on to stress that although these dualities run counter to the established tendencies of social theory before the 1980s, their essential position at the core of Giddens's structurationist thesis have been broadly accepted by human geographers in the 1980s with little critical questioning. Moreover, she backs up this claim with extensive reference to the writings of the human geography 'establishment of social theory' during the first half of the decade. We have already suggested that key human geographers such as Gregory, Pred and Thrift were involved in a positive engagement with, and 'translation' of, Giddens's sociological theory for use by geographers. Andrew Sayer, who has been a leading figure in developing realist philosophy for geographers (see Chapter 5) also seemed to be 'sold' on the ideas of duality. In a review of Giddens's 1981 text, Sayer (1983, p. 109) writes: '[The] theory of structuration . . . in my view effectively resolve[s] the problem of structure and agency, and the associated poles of determinism and voluntarism. . . . Thus, the twin errors of voluntarism (actors act independently of any constraints) and structuralist determinism (the conditions do the acting) are avoided'. Gregson also notes a series of other authors (Forbes, 1984; Johnston and Claval, 1984; Philo, 1984; Pred, 1981, 1983, 1984a, 1984b; Soja, 1985) who she claims acknowledge gaps and deficiencies in structuration theory, but who ultimately accept its merits rather than rejecting it because of its weaknesses. She makes a persuasive case for much of human geography being 'sold' on structuration as an appropriate critical approach to the subject. However, it might equally be argued that many different geographers – including some of those listed above – have effectively debated (on both theoretical and substantive terrains) the dualisms identified by Giddens, but just because they have avoided his vocabularly should not imply that geographers have not adopted a critical stance before the issues at stake.

So what exactly was it that geographers were either uncritically sold on or were fully debating? Was it the basic proposition of a duality of structure and agency, or was it Giddens's more detailed deliberations on how to establish links between them? As Gregory (1984, p. 129) noted, 'it is one thing to grasp what the theory of structuration entails . . . but quite another to incorporate its theorums into substantive accounts in such a way that one captures the engagements between agency and structure which are the fulcrum of the whole formulation'. In order to understand these questions, let alone answer them, we must delve a little deeper into Giddens's conception of agency, structure and the links between them. For, as Thompson (1984, pp. 148–9) suggests, 'What must be grasped is not how structure determines action or how a combination of actions make up structure, but rather how action is structured in everyday contexts and how the structured features of action are, by the very performance of an action, thereby reproduced'.

Agency, acts and action

Human agency may simply be understood as the capabilities possessed by people and their related activities or 'behaviours'. Giddens's theorisation in this respect begins with a differentation between

acts: that is, discrete sequences of action; and
action: that is, a continuous flow of involvements by different and autonomous human agents.

Much of the previous thinking about agency tended to refer to individual behaviour and action as a series of logical steps: 'I think, therefore I do'. Giddens, though, insists that not all action is characterised by this *purposeful* definition: that is, not all action is guided by clearly thought-out purposes the individual concerned has in his or her mind at the time the action takes place. Instead, Giddens suggests that much of the action of interest to social scientists is *purposive*, by which he means that action can be motivated by the individual who is liable constantly to examine what he or she is doing and the circumstances in which he or she is doing it. This process of self-examination Giddens calls *reflexive monitoring*, and it is this reflexive ability that permits individuals to account for, explain and rationalise their actions, either to themselves or to others.

The key question here is how this purposive action is *motivated*. Giddens acknowledges that motivation can be conscious, subconscious or unconscious, but suggests that unconscious motivation tends to be a very significant element of human conduct. Therefore, much of an individual's knowledge about the world he or she lives in is unconscious, and is not articulated as such. Giddens (1984, p. 6) notes that 'while competent actors can nearly always report discursively about their intentions in, and reasons for, acting as they do, they cannot necessarily do so of their motives. Unconscious motivation is a significant feature of human conduct'. We should note in passing, though, that it is difficult to conceive of monitoring in the case of unconscious motivations for action, given that monitoring appears to require the unconscious to become conscious.

It is in the confluence of these propositions about human agency, acts and action that Giddens prepares a model of action with which to trace the inter-connections with structure. He calls this model the 'stratification model of action' (see Figure 4.2). The model summarises these various aspects of action:

1. the reflexive monitoring of action,
2. the rationalisation of action, and
3. the motivation for action,

but also recognises that these aspects – and hence the *accounts* individuals are able to give of their actions will be affected by both the *unintended consequences* of their action and the *unacknowledged conditions* of their actions (these unacknowledged conditions include the unconscious sources of motivation discussed above). It is important to note that both the unintended consequences and the unacknowledged conditions are collective phenomena, refering therefore to the idea of action (as a continual flow of interventions in the world by different agents) rather than acts (the discrete segments performed by individual actors). Giddens suggests that this analysis exposes the limitations of other more interpretative social theories that focus on individual agents, and thereby play down these collective factors. This argument can be turned round, however, in the assertion that Giddens's stratification model of action itself misses out on some types of behaviour because it tends to polarise agency as either expressed through intentional action or through action that is unintentional or simply reactive. Moreover, these criticisms can be extended (see, for example, Philo, 1984) to suggest that Giddens effectively collapses agency into action and thereby undervalues some of the useful ideas about *thought*-and-action presented by Olsson (1980) and by Dallmayr (1982), the latter of whom argues that there is more to human agency – and particularly the more 'existential' dimensions of 'being', 'experiencing' and 'suffering' (see Chapter 3) – than Giddens usually allows.

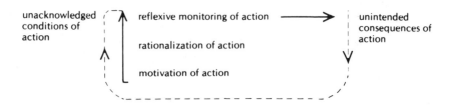

Figure 4.2   The stratification model of action (Giddens, 1984, p. 5)

Regardless of these criticisms, Giddens's model of agency begins to shed light on the nature of interconnectivity with structure. Crucially, the unintended consequences of action are seen to feed back into the unacknowledged conditions of action, for example, as with the idea of a 'poverty' or 'deprivation cycle'. Here material deprivation can lead to poor schooling, poor schooling to low-paid and low-status employment, and in turn such employment leads to further material deprivation. Both reflexive and motivated action, alongside the unintended and unacknowledged aspects of action, form important building blocks in the interpretation of interconnections between agency and structure, and it is to Giddens's view of structure we turn next.

System and structure

'[S]tructure' is usually understood by functionalists – and, indeed, by the vast majority of social analysts – as some kind of 'patterning' of social relations or social phenomena. This is often naively conceived of in terms of visual imagery, akin to the skeleton or morphology of an organism or to the girders of a building. Such conceptions are closely connected to the dualism of subject and social object: structure here appears as 'external' to human action, as a source of constraint on the free initiative of the independently constituted subject.

(Giddens, 1984, p. 16)

Giddens's notion that we have a naïve understanding of structures in society is central to the potential importance of structuration theory. Instead of vague patterning, independent of, but acting as a restraint on, human behaviour, he argues for a view of structures that

1. permits interconnectivity with human activity – action affects structure as well as structure affecting action; and
2. permits the identification of the nature and characteristics of structures.

His starting point, philosphically, is to differentiate between what he calls *structures* and *systems of interaction* in the analysis of society. Consider, he argues, the difference between speech and language. Speech can be identified in a particular time and place, and usually occurs between the speaker (or human subject) and at least one other person to whom the 'speech' is addressed. Language, on the other hand, cannot be located in a particular time and place, and is neither the product of any one speaking subject nor addressed to any particular recipient. So, says Giddens, a system of interaction in society is like speech, in that it occurs in and through the activities of individual agents, while structure (by contrast) is like language, being constituted beyond specific times and places and not restricted to the interaction of specific individuals.

Following on from this distinction, Giddens attempts to identify structure in terms of *rules* and *resources*. Resources here refer to the power invested in authority and property to enable and constrain certain social interactions through the exercise of control over people and over the material world. Day-to-day living is structured by resources, but in turn reproduces the distribution of these resources. The notion of rules is more difficult. Such rules do not refer directly to the rules we tend readily to recognise in everyday living – the rule of law; the rules of the game – but are defined by Giddens (1984, p. 21) as 'social rules':

as techniques or generalisable procedures applied in the enactment/reproduction of social practices. Formulated rules – those that are given verbal expression as canons of law, bureaucratic rules, rules of the game and so on – are thus codified interpretations of rules rather than rules as such. They should be taken not as exemplifying rules in general but as specific types of formulated rules, which, by virtue of their overt formulation, take on various specific qualities.

Giddens's use of rules, then, denotes a sequence of processes which together comprise structure:

1. *Social rules are implemented in interaction*: that is, certain regularities are consistently acknowledged in day-to-day encounters, as people have a knowledgeable yet sometimes unconscious scheme of what behaviour is and is not appropriate in particular situations. Such knowledge of rules is not in the form of a 'handbook' of responses to all possible social encounters. Rather it represents a 'generalised' capacity to respond to and exert influence on a wide range of potential social circumstances.
2. *Social rules structure interaction*: that is, it is these rules that produce, sustain, finish and reproduce social encounters.
3. *The rules that structure interaction are themselves reproduced by the processes of interaction*: that is, the rules themselves can be modified or reformulated during the processes and procedures of interaction.

It follows from these three propositions that the 'structural properties of social systems', as formulated in the rules and resources postulated by Giddens, are both a *medium* of social practices and an *end result* of social practices. Here, then, is the essential duality of structure worked out in more detail. Thus far we have considered rules and resources at the level of broad interactions in society. It is important to note, however, that Giddens (*ibid.* p. 24) focuses particularly on the rules and resources of structure positioned within social *institutions*: 'The most important aspects of structure are rules and resources recursively involved in institutions. Institutions by definition are the more enduring features of social life. In speaking of the structural properties of social systems I mean their institutionalised features, giving "solidity" across time and space'. It is these very institutions that have previously been portrayed as 'barriers' to social action, yet Giddens here argues that structure (in the sense of solid enduring patterns) is not a barrier to action – as posited within some more functionalist approaches – but is essentially embroiled in the production and reproduction of action. It is this notion of structure involving both regularised patterns and the duality of interaction, which is carried forward to a particular theorising of the interconnectivity of agency and structure.

Giddens's stratification model

Figure 4.3 represents Gregory's (1986) augmented version of Giddens's 'stratification model of structure', which attempts to present an interwoven analysis of the rules and resources that comprise structure. Simply explained, the model traces the links between systems of interaction (relations produced in and through the actions of individual agents, like speech) and structure (relations not restricted to particular subjects, times and places, like language). He identifies these links in terms of lines of mediation or *modalities* between interaction, on the top line of the diagram, and structure, on the bottom line. Three dimensions of modality are suggested:

1. Where individual agents interact, they engage in a *communication* of meaning, and in so doing they draw upon interpretative schemes. These schemes of interpretation are capable of identification and analysis at the level of structure as *semantic rules*. Thus, social interaction through communication can be structured because particular interpretations of reality can be signified in our language beyond the simple meaning of mere words and thoughts, and Giddens terms this process *signification*.
2. When individual agents interact they apply, consciously, subconsiously or unconsciously *sanctions* on their behaviour. In so doing they are drawing upon social norms that can be identified and analysed at the level of structure as *moral rules*. Thus, what is legitimate, and what is not, in social interaction through sanction can be structured by this *legitimation*.
3. The use of *power* in social interaction involves recourse by individual agents to certain enabling facilities so as to secure particular outcomes from the interaction. These enabling facilities can be identified and analysed at the level of structure as *resources* that, as mentioned previously, comprise the structures of *domination*. He differentiates between those resources that produce power over people and therefore point towards the enabling factor of *authority*, and those resources producing power over objects, reflecting the *allocation* of resources such as wealth and property.

Clearly, these three dimensions or modalities are in practice blurred rather than distinct. In day-to-day life semantic rules are often bound up in moral rules and vice versa. Equally, power will be produced and reproduced through communication and moral sanction as well as through the authorisation of behaviour and allocation of resources. However, we may say with Gregory (1981, p. 10) that

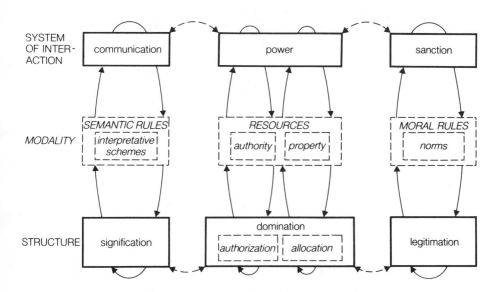

Figure 4.3   The stratification model of structure (Gregory, 1986, p. 465)

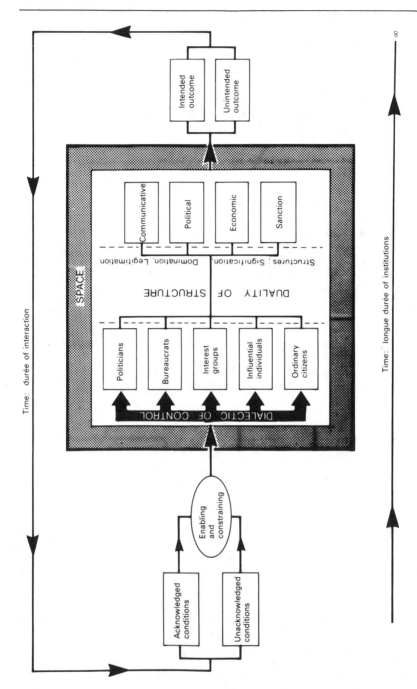

Figure 4.4   A model of the structuration of urban space (Moos and Dear, 1986, p. 245)

What is important about characterisations of this sort is not so much the systems and structures which they distinguish – where there is no doubt a lot of room for argument – but rather their recognition that in drawing upon these various modalities the actors involved are displaying some degree of penetration of practical life.

It is in this manner that Giddens's models of the stratification of agency and structure have been adapted by researchers as the basis for empirical investigation. An example of this reworking of the structuration model is given in Figure 4.4, where Moos and Dear (1986; see also Dear and Moos, 1986) work out a model for analysing the 'structuration of urban space'. Although they embellish considerably the framework provided by Giddens, the essentials of structuration – the duality of structure, the complex nature of modality, and the reproduction of action and structure – are retained.

Giddens has therefore provided us not only with a retheorising of society but also with something of a model on which to base empirical investigation. Gregory (1986) neatly summarises the qualities of how this model operates. First, it suggests to us that, because all social interaction involves communication, sanction and power, all systems of social interaction will depend on the connecting structures of signification, legitimation and domination. Second, it emphasises that social practices are inextricably linked with social structures, such that as social practices are produced so social structures are *re*produced. This viewpoint ensures that (to use Giddens's well-worn phrases) the 'personal transient encounters of daily life' are essentially bound up with the 'long-term sedimentation of social institutions'. Third, the model allows a constant potential for change with the flux of production and reproduction of social life. Conflicts between individuals within systems of social interaction, and contradictions between structural principles, ensure that structuration will avoid functionalist and determinist outcomes.

A *brief review of early criticisms of structuration*
We mentioned at the beginning of this chapter that any attempt to occupy the middle-ground theory between and interconnecting structure and agency, was bound to attract critical accusation for being an unacceptable compromise to the 'purity' of other approaches. It might be anticipated that Giddens and others of the structuration school would have been heavily criticised by theorists who prefer to cling to both voluntarist and determinist notions alike. In fact (although this should not be conflated with voluntarism and determinism), the fact than most critical comment has stemmed from the Marxist camp could suggest that structuration theory may be closer to humanist thought than it 'lets on'.

Giddens is criticised, first, on the grounds that he has been unable to transcend the two very different theories he attempts to merge in structuration theory. For example, Callinicos (1989), offering a critique of Giddens in a text called *Marxist Theory*, seeks to undermine the primary and fundamental objective of bringing together structural and interpretative sociologies (p. 145):

> [L]et us note what such an attempt to arrive at a synthesis of mainstream sociology and Marxism implies. It supposes that the interpretative sociology of Weber . . . can be rendered compatible with elements at least of historical materialism. It seems to me that there are general grounds for doubting this to be so: the two traditions are so fundamentally opposite in method, substance and political implications that the actions of any attempt to combine them can only be incoherence, eclecticism, or the denaturing of one or both.

Such sweeping refusals to contemplate middle-ground theory are common-place, and unlikely to be resolved, but there have been critics who have been more willing to discuss particular aspects of Giddens's discourse. We have noted earlier in the chapter, for example, that he has been been gently chided for collapsing agency into action (Philo, 1984) and for thereby obscuring the recognition of dimensions to agency that fall somewhere between (or maybe even outside of) the polarities of intentional ability and reactive behaviour. Generally, though, the theorising of agency within structuration has been less challenged than that of structure, perhaps because – and as Callinicos (1989, p. 145) again suggests – Giddens is seen by many commentators as continuing 'to give primacy of certain fundamental Weberian premises: the priority of sub-ject over structure, and the omnipresence of domination'.

Criticism of Giddens's formulation of structure has taken many different forms. One principal focus of argument is his idea that structure may be explained as rules and resources. Criticism here disputes the clarity of Giddens's own thought on this point ('if Giddens expects readers to accept his proposal to conceive of structure in terms of rules and resources, then the onus is on him to provide a clear and consistent examples of what he would count as a relevant rule'; see Thompson, 1984, p. 158) and also asks whether rules and resources should really be seen as distinct from social structure. Thompson, for one, argues strongly for such a distinction. He suggests that Giddens's sup-posed failure to offer a precise definition of the term 'rules' is inevitable be-cause the rules crucial to social structure cannot be clarified without presupposing a 'criterion of importance': which rules are important, for ex-ample, in analysing the social structure of capitalist societies? It is argued that the criterion of importance cannot be derived from analysing rules alone, but must make broader reference to ideologies or world views that offer advice about what is and what is not important. Thus, semantic rules may well be a relevant factor in analysing social structure, but 'to study their differentiation presupposes some framework, some structural points of reference which are not themselves *rules*, with regard to which these semantic rules are differenti-ated' (*ibid.* p. 159). Equally, Giddens's focus on rules does not appear to answer questions of how the range of options available to individuals and groups are differentially distributed and structurally circumscribed: that is, the whole question of power.

Urry (1982) approaches the same point from a different tack. He suggests that the treatment of structure in structuration does not deal adequately with the notion of socially structured struggle and its unintended effects:

> social structures are heavily interdependent and in many situations there are both profound social struggles *and* a process by which the effects of such struggles are constrained so as to produce 'reproduction' in some sense. Thus, satisfactory analysis involves both explaining why particular struggles happen to materialise, and to con-sider the particular constraint upon the consequences of these specific struggles.
>
> (Urry, 1983, p. 103)

Structuration is therefore criticised because it does not seem to cope with, for example, the complex mechanisms that occur when a particular social class or social force enters into a struggle so as to achieve certain goals such as higher wages. There seems to be a regular effect here of producing a different yet reproducible portfolio of social consequences, whereas structuration seems to assume that these structured properties should be viewed merely as the knowledgeable accomplishments of social action.

Giddens offers some response to his critics both in print (see, for example, Giddens, 1982; Gregory, 1984) and in his subsequent reformulations of structuration theory. It is to this second phase of his project we now turn.

Giddens's structurationist project: phase II

Giddens's later work on structuration theory (see Figure 4.1 with reference to his 1981, 1985a and 1989a texts) has consisted of a dual thrust: first, a 'deconstruction' of historical materialism and further engagements with Marxist theorists; and, second, a development of his theory through the inclusion of time–space concepts into structuration. The first of these developmental thrusts can be dealt with relatively briefly as it follows through many of the themes already mentioned in this chapter. Giddens saw the need throughout the 1980s to engage with Marxist theory, raising substantive criticisms of it and coterminously – but not coincidentally – highlighting the advantages of his own structurationist thesis in the process. In these works, then, can be found further debates over the functionalism of historical materialism (structuration being explicitly non-functionalist), its evolutionary nature (structuration being non-evolutionary) and its preoccupation with forces of production as the engine of change in society. On this latter point, Giddens serves up illustrations of non-capitalist societies to suggest that the forces of production are not paramount in these cases, and also points to the role of the nation–state as a means of exercising violence through military strength. This analysis again acts to posit an important facet of social change that is not central to the historical materialist thesis.

As we show in Chapter 2, many of these arguments have been part of the currency of the continuing Marxist discourse in social science during the 1970s and 1980s, prompting some commentators both to question Giddens's reading of contemporary Marxist theoretical development and to wonder whether in fact Giddens had not been moving in roughly the same direction as much recent Marxian social science (see Callinicos, 1989; Gore, 1983; Wright, 1983). Giddens himself makes clear statements about the relationship he see between his work and Marx's writings:

> I don't regard myself as a Marxist and have never thought of myself as a Marxist. I've always thought of what I do as using important aspects of Marx's writings and also important aspects of subsequent Marxist thinking, but in a highly critical way, and I would resist the idea that there's anything more than a generalised appropriation of the notion of *praxis* from Marx in structuration theory.
>
> (In Gregory, 1984, p. 127)

And it would appear that in this debate between Giddens's structurationist thesis and contemporary Marxist thinking, there is ample scope for a prolonged engagement!

Giddens and time–space theory

Time and space had been key organising principles in the early formulation of structuration, but the second notable theme of Giddens's later work is that of developing these notions of time and space so that they may become integral facets of structuration theorising. In his 1981 and 1984 texts Giddens draws on the work of particular human geographers (Hägerstand, 1975; and those who have developed and applied his ideas, for example, Gregory, 1978; Pred, 1977, 1978; Parkes and Thrift, 1980; Thrift and Pred, 1981), and so it is important for us to follow Giddens's thinking here by outlining the significance of the links he detects between *time-geography* and structuration theory, and by then taking note of the specific criticisms which he raises against the work of time-geographers.

It is often argued that time-geography as developed by Torsten Hägerstrand provides a 'language' with which to capture the interactions of agency and structure central to Giddens's formulation of structuration theory. And this is particularly the case given that Giddens insists on recognising the grounding of such interactions in everyday 'time–space settings': in other words, at certain moments in certain locations. The first concern of time-geography is with the *agency* (and note the echo here and elsewhere of structurationist terms) of individual and 'indivisible' human beings, and the stress is immediately on the indisputable fact that in order to 'act' – to eat, to work, to find partners – people have to move through time and across space. A typical time-geographic diagram (Figure 4.5) hence represents the daily movements an individual makes through time and space, from getting up in the morning at one location (home) through to his or her return to this location after working, shopping and socialising. And it is possible to imagine such a diagram drawn for one individual's movements over much longer time-spans, of course, or drawn to show the simultaneous and often overlapping movements of many different individuals. There is a complex vocabulary now built up to describe the human world in time-geographic terms, and this is the vocabulary of 'paths',

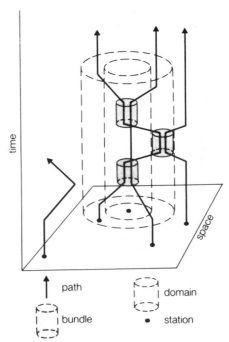

Figure 4.5   Hägerstrand's web model of time-geography (Gregory, 1986, p. 485)

'projects', 'bundles', 'coupling', 'time-budgets' and so on (see Thrift, 1977; Parkes and Thrift, 1980; Carlstein, 1981; Pred, 1977, 1981), but the chief importance of these diagrams and their associated vocabulary lies in their depiction of regularities in how individuals *repeatedly* draw upon – and in how different individuals *simultaneously* draw upon – the *resources of time and space*. These regularities amount to 'time–space structures' deeply engrained in the conduct of everyday life, and they themselves are embedded within (are shaped by; are constituted by) a more intangible realm of *structure*: the realm of economic, social, political and cultural structures such as those governing the distribution of wealth and power in modern capitalist societies.

Time-geography thus allows us both to conceptualise (but in a more tangible fashion than does structuration theory alone) and to represent for the lives of particular people the meeting of agency and structure. It allows us to keep one eye on the agency of individuals moving about in time and space, and to keep the other on the time–space structures of their lives – their regularised movements between locations and along particular routes – as governed by their economic, social, political and cultural circumstances. Moreover, a time-geographic perspective allows us to see that *creative* acts on the part of human beings – their decisions, particularly when taken collectively, to alter their movements in time and space (perhaps by relocating workplaces, by redesigning cities, by re-scheduling transactions) – can on occasion *transform* the structures in which they exist. The lens of time-geography hence enables us to pick up the *recursive* interaction of agency and structure, as 'played out' through the media of time and space, and thereby to visualise what Giddens is asking us to understand about the constitution of social life.

Giddens himself realised early on that structuration theory demanded precisely this attention to the 'timing' and the 'spacing' of how 'agency influences structure influences agency *ad infinitum*' (see his Chapter 6 on 'Time, space, social change' in Giddens, 1979) and, in some of his later writings (Giddens, 1984, Chapter 3; 1985b), he reflects further upon such matters. In so doing he draws (albeit critically: see below) upon the literature of time-geography, and so it is not just a case of geographers bringing Giddens's arguments into geography: there is also an element of sociologists bringing time-geographic arguments to the attention of the social sciences more generally.

An excellent example of how a time-geographic perspective throws light on questions of agency and structure can be found in Jacqueline Tivers's work on the constraints experienced by young mothers in the course of their daily lives (see Tivers, 1985, 1988). Tivers used time-geographic tools to depict the complex 'juggling act' in time and space performed by 400 women in the London borough of Merton: their careful management of when and where they performed tasks such as housework, shopping, child-minding, picking up children from nurseries and schools, often without access to a car and often at the whim of public transport. She found that their lives tended to be spatially localised, in that their range of movement was severly limited by *physical constraints* arising from the arrangement of city spaces (land uses, buildings, roads). She then argued that these constraints must be thought of as being 'themselves the products of societal structure' – as being the outward expression of more fundamental *social constraints* to do with the divisions, relations and conflicts present in a society's functioning – and more specifically she claimed that the Merton women suffered from *gender-role constraints* bound up with deep-seated assumptions about the different roles 'naturally' performed by women and men, with women consigned to 'childcare' and men to 'providing', that fed into an acute spatial separation of homeplace and workplace (a separation that had numerous knock-on effects for where shops, schools, surgeries and so on were located). Tivers thus demonstrated compellingly how the agency of the Merton women was very directly shaped by the sedimented structure of gender relations, as intimately bound up with the time-geography of their everyday lives.

However, it should also be emphasised that Giddens himself is critical of time-geographers in general and Hägerstrand in particular, given what he considers to be a 'naïve and defective' conception of the human agent in their work. Four main criticisms are advanced in this connection (Giddens, 1984, pp. 116–19):

1. Hägerstrand is accused of treating individuals as somehow 'constituted' separately from the social settings of their day-to-day lives (they are taken as 'given', rather than seen as human subjects whose very subjectivities are fashioned in the course of day-to-day living and working).
2. It is suggested that Hägerstrand and others thereby tend to 'recapitulate the dualism of action and structure', and in so doing under-emphasise the transformational nature of human action.
3. Hägerstrand uses the idea of constraints in his work, focusing particularly on the constraining properties of the human body. Giddens sees all constraints as potential opportunities, though, and reacts against the notion of constraint by suggesting that there is more than a hint of historical materialism mixed in with time-geography.
4. Time-geography suffers from a 'weakly developed theory of power'.

Following the now 'customary' pattern of critique then theory-building, Giddens picks over the meaty morsels of truth in time-geography and then proposes some fundamental propositions about how time and space are placed theoretically within structuration. One of his initial premises is that social systems are not only structured by rules and resources but that they are also situated within time and space. It follows that considerations of time and space will mark the boundaries of social analysis as well as constituting the frameworks within which social life takes place. Moreover, social interaction will represent a mix of presences and absences in time and space. Some interactions take the form of face-to-face contact where both actors are present at the same time and in the same place. In other circumstances social interaction occurs where the other actor is not present in the same place at the same time. The progress first of writing, then of transport, media and telecommunications have transformed the time–space of social interaction, such that an individual in New Zealand can interact with another in Europe within even days or so via air-mail letter, typically with only momentary delay by telephone (and this is changing now) and all but instantaneously with contemporary telecommunication equipment.

These interactions where the other is not present in time and space are implicated in what Giddens terms *time–space distanciation*, and he argues strongly that both agency and structure should be conceived not only as they have been described already in this chapter but also in terms of this distanciation: that is, according to how social systems are integrated over time and space. It is in order to illustrate the potentially different levels of such integration that Giddens differentiates between *social integration* and *system integration*. Social integration, he argues, is the day-to-day routine of life where actors are usually co-present in time and space. So far as system integration is concerned, however, Giddens notes that similar social practices occur in different times and places. Such practices can be traced back into the structures that have been sedimented in previous times and/or other spaces. Analysis of system integration therefore refers to interactions that are constituted outside of current time and space (the 'here and now') and involve some element of interconnectivity with others who are separated (are distant) in time and space.

Later versions of Giddens's structuration project therefore aim to demonstrate how social activity in current time–space (that is routine interactions with people who are present) is affected by other social relations that do not belong to current time–space. In this way, time–space is transcended by the 'stretching' of social relations over time and space: that is, by time–space distanciation. He uses the components of the stratification model of structure (see Figure 4.3) to illustrate how time–space distanciation is 'transported' through the interconnections between structure and agency:

> The underlying thread of my argument is as follows. Power is generated by the transformation/mediation relations inherent in the allocative and authoritative resources comprised in structures of domination. These two types of resource may be connected in different ways in different forms of society. It is certainly a mistake to suggest . . . that the accumulation of allocative resources is the driving principle of all major processes of societal change. On the contrary, in non-capitalist societies it seems generally to be the case that the co-ordination of authoritative resources is the more fundamental lever of change. This is because . . . authoritative resources are the prime carriers of time–space distanciation.
>
> (Giddens, 1981, p. 92)

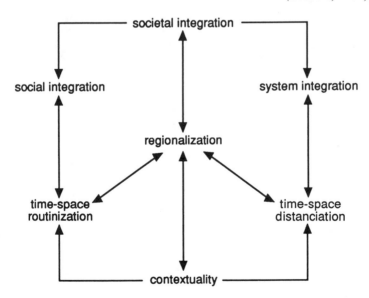

Figure 4.6 A model of the structuration of time–space relations (Gregory, 1986, p. 467)

Thus in small human societies such as bands of hunter-gatherers, so he argues, extension of time–space organisation is limited to the spatial extent of foraging trips and the perambulations of the group as a whole, and to the temporal extent of tradition and kinship rules. The level of time–space distanciation characteristic of such societies is therefore low, with few regularised transactions with others who are absent in space as would be the case in larger, more highly technological societies.

Building upon these arguments Giddens produces a model of time–space relations given in Figure 4.6, again in the form of Gregory's (1986) augmented version. The model traces the two possible routeways of societal integration:

- *Social integration*, which reflects the time–space routinisation of agents living, working, playing and so on together in time and space.
- *System integration*, which allows for variable time–space distanciation with agents not present in current time and space.

Furthermore, Giddens argues that social integration and system integration come together in particular *locales* (a term he uses to mean something more than just the 'stage' of action: rather, for Giddens locale is an integral part of the constitution of social action). Thus, routine interaction and distanciated interaction meet in what he terms *modes of regionalisation*, which channel and, in turn, are channelled, by the pathways of time–space followed by both the day-to-day activities of individual actors and the institutions of social systems. Regionalisation hence refers to the 'zoning of time–space in relation to different forms of interaction, and therefore neccessarily occurs at various scales as well as in different modes' (see Figure 4.7). For example, Giddens talks about the social gathering as regionalisation. Here, the *form* or boundaries of regionalisation will be indicated not so much by physical barriers as by (for instance) body posture and positioning (see Seamon's work on 'time–space ballets' – see Chapter 3). Such encounters will be of short *duration* and will occur in a narrow *span* of time–space. On a much grander scale, Giddens also envisages regionalisation that incorporates zones of considerable time–space span and far-flung form, over which interaction occurs for a much longer

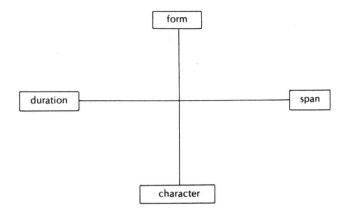

Figure 4.7   A classification of modes of regionalisation (Giddens, 1984, p. 121)

duration. Regionalisation is not necessarily a reference to geographical region, then, but an expression of the 'structuration of social conduct across time–space'. As such it will vary in *character* according to the way in which localised time–space organisation is ordered within more deeply sedimented social systems:

> Thus in many societies 'the home', the dwelling, has been the physical focus of family relationships and also of production, carried on either in parts of the dwelling itself or in closely adjoining gardens or plots of land. The development of modern capitalism, however, brings about a differentiation between the home and the workplace, this differentiation having considerable implications for the overall organisation of production systems and other major institutional features of contemporary societies.
>
> (Giddens, 1984, p. 122)

The form, duration, span and character of regionalisation helps make sense of how time–space routine and time–space distanciation come together. Through regionalisation, Giddens claims, theory can capture the flow of human agency as a series of situated events in time and space: what he calls *contextuality*.

Space and scale in regionalisation

In some of his later work, Giddens (1984, 1985a, 1985b) has focused on the way in which regionalisation characterises particular places. One of the major criticisms of his thinking in the second phase of theorising structuration is that he has been unsuccessful in specifying an hierarchy of locales arising from structuration in society. Commentators such as Thrift (1985) suggest that he has been preoccupied by either the micro-scale or the macro-scale, and has thereby ignored potential hierarchical levels in between. In fact, Giddens strongly rejects this criticism, not on geographical grounds, but because he sees micro- and macro-studies as bound up in the debate over the duality of structure and agency. Thus he chooses not to employ the terms macro- and micro- for two reasons:

1. Certainly in sociology, the two are often set off against one another in the sense that one is a better scale of approach than the other. Such confrontations are phoney, argues Giddens, because structuration allows us to understand the interconnectivity of the micro-worlds of face-to-face interaction and the macro-worlds of systems subject to time–space distanciation.
2. Even if such confrontations are avoided, there tends to be what he regards as an unfortunate division of labour wherein micro-scale studies concentrate on the activities of the human 'free' agent whereas macro-scale studies tend to analyse the structural constraints that bound that activity.

Typically, then, Giddens turns the argument around to his own favour by reverting to the core theme of duality. The study of scale over space necessarily in his view takes us away from the core interconnectivities between social interaction and system interaction:

> the spatial differentiation of the micro- and macro- becomes imprecise once we start to examine it. For the forming and reforming of encounters necessarily occurs across broader tracts of space than that involved in immediate contexts of face-to-face interaction. The paths traced by individuals in the course of the day break off some contacts by moving spatially to form others, which in turn are broken off and so on. . . . It is apparent that what is being talked about under the heading of micro/ macro processes is the positioning of the body in time-space, the nature of the

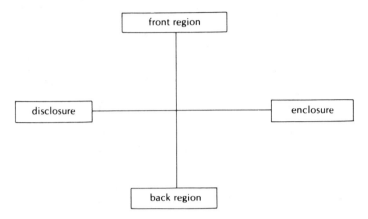

Figure 4.8   Regionalisation and zoning (Giddens, 1984, p. 124)

interaction in situations of co-presence, and the connection between these and 'absent' influences relevant to the characterisation and explanation of social conduct.

(Giddens, 1985b, p. 292)

Despite this reflex reversion to the social roots of structuration when questioned on matters of scale and space, Giddens has in fact developed the notion of regionalisation in such a way as to parallel the concerns and interests of many human geographers. For example, he suggests that it is not only the co-presence of actors (see Figure 4.6) but also the *presence-availability* – that is, the potential for actors to be able to 'come together' – which affects the scale and type of regionalisation. Co-presence and presence-availability help us to understand how regionalisation is accomplished in different settings. Just as 'facing' and 'turning your back on' are integral to body-positioning in normal social encounters, so Giddens argues that regionalisation bounds in zones of time–space that are characterised by sustained interactions between *front* and *back* regions (see Figure 4.8). Furthermore, 'frontness' and 'backness' are associated with, but not coincident with, the idea of *disclosure* (the display of particular characteristics) and *enclosure* (the covering up of those characteristics). As Giddens (1985b, p. 277) notes, 'these two axes of regionalisation operate in the complicated nexus of possible relations between meaning, norms and power'. Here, of course, he mirrors (and some would say mishandles) pre-existing spatial concepts such as Goffman's front-and-back regions.

The integration of power into regionalisation is crucial to the conceptualisation. Front regions, where interaction is open, 'on show' and often official, are related to particular rules and resources. Here, surveillance is an important agent of supervising interaction at many different levels: the modern capitalist state moblising laws that govern criminal behaviour, tax gathering, and political change; the workplace setting where productivity, timing and performance are monitored; the dinner party where dress, conversation and choice of menu will be affected by perceived codes of practice. Back regions (note here the familiarity of this terminology in terms like 'backwater', 'backwoods', 'backstreet') can also be seen at different scales of interaction: areas of privacy or solitude; areas of 'underdevelopment'; the tea-room or toilet in the workplace; the garden shed or other corner of solitude in the home. Giddens uses the

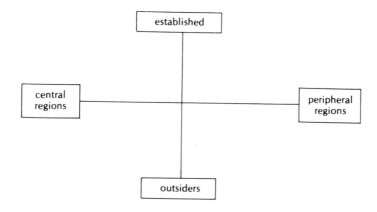

Figure 4.9 Centre-periphery distinctions (Giddens, 1984, p. 131)

example of urban neighbourhoods to illustrate the differentiations between enclosure, disclosure, back and front. Industrial neighbourhoods were once proudly and visibly displayed in cities, but now they tend to be treated as unsightly back regions, hidden away as 'enclosed enclaves' or relocated to the edge of the city (see also the relevant discussion in Chapter 3). Equally, the back areas of decaying docklands can be transformed into front areas through planned gentrification (see Chapter 6).

A further aspect of regionalisation occurs in notions of *centre* and *periphery*, which Giddens equates with those who *establish* themselves as having control over resources, and those who are *outside* that control (see Figure 4.9). Although geographers have tended to use 'core-periphery models' across large spans of time and space (Dodgshon, 1987, for example, in his treatment of the historical geography of state formation provides a lengthy analysis of cores and peripheries shifting across large tracts of time and space), it is not difficult to denote cores and peripheries at other scales: in the world economy, the nation state, the region, the city, the workplace, the home and so on (see N. Smith, 1984). At each of these scales the 'established' (nation, capital fraction, class fraction, bosses, males who perceive themselves to be leaders of households) employ a range of social tactics to distance themselves from others who are treated as inferior or outsiders. Each of these relationships changes over time, a point illustrated by Giddens in terms of the interconnectivity of city and countryside. Using the work of Mumford (1960), he views the traditional city as a 'container' for the generation of power through structures of authorisation and allocation. With the passage of time, however, the commodification of space brings about new forms of institutional order, which have changed the circumstances of both social and system integration, and thereby reduced the remoteness of the countryside in both time and space.

These treatments of scale and space in structuration are of obvious potential relevance to human geographers in their wish to understand the relationship between society and space, and between structure and agency (see, for example, the parallels in the work of Sack, 1980; N. Smith, 1984; Taylor, 1989). Ultimately, however, the work of Giddens and others in the structurationist school will be evaluated by most human geographers in terms of the ability to inform in practical and direct terms the particular foci of their study. It will therefore be insufficient merely to offer a critique of Giddens's theorising on regionalisation *per se* without reference to key individuals who have sought to 'translate' structuration into the geographical arena. Indeed, we would want to stress at this point that geographers have not merely been slaves to Giddens's structuration theory, but rather have worked with particular strands of it and have perhaps even improved on them. Therefore we now devote most of the last part of this chapter to discussing some of the work conducted by geographers who have sought to enlarge the treatment given in structuration theory to the geographical terrain of time and space. In this context we would certainly want to acknowledge the contribution of Nigel Thrift, who is currently Professor of Geography at the University of Bristol, since Thrift has been one of the key 'translators' of structuration theory for geographers (see, for example, his seminal 1983 paper) and has also been influential in providing empirical interrogations of structuration theory (see, in particular, his 1985 essay). But in the limited space remaining to us here we have decided to restrict our focus to two other influential human geographers: the first, Allan Pred, has provided us with structurationist studies of place and of biography formation; and the second, Derek Gregory, has challenged and refined structurationist concepts to deal with what he calls 'the historical geography of modernity'.

Structuration in human geography: two examples

Allan Pred: the 'becoming' of place and biography formation
Allan Pred is currently Professor of Geography at the University of California at Berkeley, and his contributions to the debates on time-geography and (more latterly) on the place of structuration in the theoretical baggage available to geographers have been substantial. Although more latterly (see Pred, 1990) he has suggested ways in which Giddens's conceptualisations may be superceded (certainly to the extent of avoiding use of the term structuration), much of his work in the mid-1980s reads as a direct engagement with structuration theory. Applying it to the changing nature of places in the Swedish province of Skåne (Pred, 1985; 1987) led him to express the notion of the 'becoming' of a place: '[P]rocess-participants are regarded as integrated human beings who are at once objects and subjects and whose thoughts, actions, experiences and ascriptions of meaning are constantly *becoming*, through their involvement in the workings of society and its structural components as they express themselves in the becoming of places' (Pred, 1985, p. 338); and he summarises his theoretical involvement with structuration in a series of propositions that describe place as 'an historically contingent process', and that are clearly linked to the discussion of structuration theory (as well as of time-geography) earlier in the chapter:

1. Structuration is continuous, and the interconnecting processes of social reproduction and individual socialisation are spelled out by the intersecting paths of individual action and institutional activity in particular times and spaces (see the sections on *structure* and *agency* above).
2. What happens in a place, and the meanings attached to that place, is inseparable from the structuration process in that place *and elsewhere* (see the section on *time–space distanciation* above).

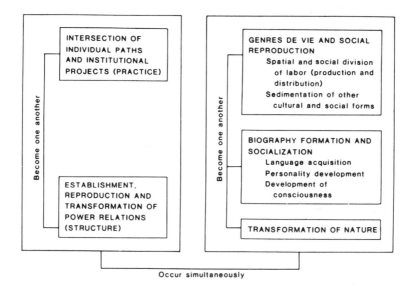

Figure 4.10 Components of place (and region) as historically contingent process (Pred, 1985, p. 343)

3. Power relations are at the heart of the social structure of a place as it becomes, but can themselves be transformed in the process (confirming the *duality of structure* as discussed above).
4. Power relations also influence what people know, what they are able to say about what they are doing, and how they perceive and think (through the *reflexive monitoring* of action discussed above).
5. The generation, regeneration and transformation of power relations are essentially intertwined with the becoming of a place, and the manner of this intertwining depends on the degree to which localised institutions are controlled by non-local system integration rather than face-to-face social integration (and here the link is with Giddens's *modes of regionalisation* as discussed above).
6. The becoming of place is also interwoven with individual biographies.
7. The role of institutional projects is dominant in the becoming of a place. Projects relating to production or distribution involve a *spatial* division of labour (as they are not geographically ubiquitous) and a *social* division of labour (as participation levels vary). These dimensions also refer back directly to *time–space distanciation*.

Pred (1985, p. 344) suggests that (as in Figure 4.10)

> Taken together these overlapping positions mean that any place or region expresses a process whereby the reproduction of social and cultural forms, the formation of biographies, and the transformation of nature and space ceaselessly become one another at the same time that power relations and time-specific path-project intersections continuously become one another in ways that are not subject to universal laws, but vary with historical circumstances.

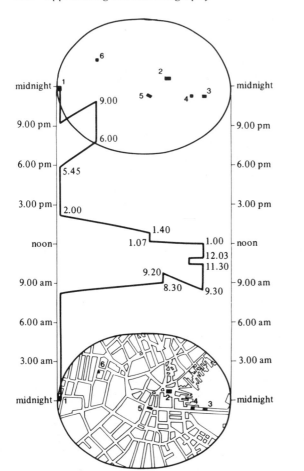

1 home
2 Faneuil Hall
3 countinghouse
4 coffeehouse
5 'change' on State Street
6 club-meeting site

Figure 4.11   Daily path of a typical Boston merchant capitalist (Pred, 1984b, p. 268)

Pred's (1984b) work on mercantile capitalists in Boston during the late eighteenth and early nineteenth centuries includes a useful example of how both the time–space specifics and details 'as experienced' of biography formation can be drawn out in a structurationist analysis. Figure 4.11 represents a view of the daily path of a 'typical' merchant. He would arise sometime between 6.00 a.m. and 8.00 a.m., dress, eat a hearty breakfast and then leave his home in a wealthy residential area to be accompanied by his servant to the Faneuil Hall market area where he would purchase the day's provisons for dinner (fish, meat, cheese, fruit and vegetables). After unhurried conversation with friends at the market he would proceed to his 'countinghouse' workplace, sited in the harbourside area of wharves. For the next three or four hours he would attend to his correspondence, his accounting and other management tasks, interrupted by a short trip to a nearby coffeehouse or tavern. At 1.00 p.m. he would walk from his 'countinghouse' to the 'change' (or exchange) where all of the mercantile capitalists of the city would conduct informal business meetings. By 2.30 p.m. it was time to return home for dinner (often with merchant friends) followed by perhaps a nap induced by prandial refreshment. In the evening he might go to one of the numerous 'small gentlemen's clubs', where he would again encounter other members of the capitalist merchant class, before arriving home again around midnight.

The purpose and usefulness of such details in the context of structuration is clearly argued by Pred (*ibid.* p. 269):

> shared daily path elements and repreated interactions at Faneuil Hall, coffeehouses, 'change', dinners, and clubs facilitated the acquisition of the more or less confidential knowledge necessary to the definition and execution of mercantile projects. Moreover, the daily, reproduction of these institutionalized forms of 'sociability' (Thrift 1983) resulted in the personal accumulation of common knowledge, or *in a deepening of collective consciousness*, and a further entrenchment of the basic stances of Federalist ideology.

Thus his identification of shared elements of daily paths in the biographies of particular individuals leads to a recognition of social interactions, and more specifically of exchanges of knowledge, which would sponsor a form of political mobilisation whenever the interests of mercantile classes were threatened by micro-scale or macro-scale events. Pred argues that such political mobilisation in turn can produce a restructuring of the explicit rules that operate at local, state or national levels.

There can be little doubt that Pred's engagement with structuration has been influential in the wider engagement with structuration by human geographers. The question arises, however, of the degree to which structuration can enhance the practice of geographical understanding. Pred is able to conclude his study of villages in the Swedish province of Skåne by stressing that concepts derived from structuration do usefully present themselves to geographical research: the fine-grained details and biography formations did vary from place to place; path–project intersections were given unique qualities by local practices and power relations, which had been sedimented over a long period of time; agency, and the 'internal–external' dialectic were important in each place; there was significance in the languages used by residents of different villages, with consequences for how action was enabled and legitimised; and so on. Such work has been met by two polarised attitudes. Either it has been accepted that the deployment of structuration theory has sensitised geographical analysis and permitted a rethinking of human geographies as found 'on the ground', or it has been objected that the research carried out in the name of structuration theory appears to bear a remarkable resemblance to other such work under different theoretical headings, the main differences being the categorical boxes in which findings are presented and the transformation of the vocabularies of analysis and interpretation. Pred's influential translation of structuration theory, while not being uncritical, has perhaps represented more of a seminal *application* of the wisdoms of the structurationist school than a sustained *development* of them in the geographical arena (though see Pred, 1990).

Derek Gregory: advancing structuration in human geography
Derek Gregory is currently Professor of Geography at the University of British Columbia at Vancouver, after having worked previously at the University of Cambridge. He is beyond doubt one of the most influential and formative

contemporary thinkers in human geography, and his careful and complex treatment of Giddens's structuration theory has taken the form of peeling off layer after layer and of subjecting each to careful scrutiny. There is insufficient space here to offer more than a brief summary of the complex deliberations presented by Gregory, and detailed reference should be made to his own writings in critique and extension of Giddens and structuration theory (see Gregory, 1981, 1982c, 1985b, 1989a, 1989b, 1989c, 1990b). Equally, there appears to be no one starting point for the critique. Gregory is able to take any one of a number of outer layers of structuration and cut his way through towards the deeper interconnectivities in true critical fashion. We should emphasise at the outset, however, his repreated stress that 'I have learned a great deal from Giddens's writings. If I am sceptical about some of the claims that have been made about structuration theory – by advocates as much as by critics – that does not mean that I am in any sense dismissive of its importance' (Gregory 1989b, p. 213). In this context we will mention two trajectories of Gregory's engagement with structuration. First, we describe his use of structurationist concepts in a study of the historical geography of the Yorkshire woollen industry, and then we investigate his analysis of the need to include what he sees as the full implications of time–space theory in structuration concepts.

Derek Gregory was perhaps the first to read Giddens's discriminations between 'structure-orientated' and 'agency-orientated' social theories into the recent history of human geography (see especially Gregory, 1981), and in so doing to suggest that Giddens's attempt to bridge the structure–agency 'gap' – namely, his development of structuration theory – could signpost a way forward for human geographers looking

1. to transcend the opposition between Marxist and humanistic geographies; and
2. to take seriously the role of time and space in the 'constitution' of social life.

Moreover, relatively early on in his grappling with Giddens's structurationist materials Gregory sought, chiefly in his *Regional Transformation and Industrial Revolution* (1982b), to mobilise such materials in a substantive work on the historical geography of the Yorkshire woollen industry. Gregory (pers. comm.) once said that he would have preferred the book to be called *Conscripted into Mills*, and this title would have captured his empirical concern for the transformation – as this worked itself out in specific fashion in a specific 'bounded' region – whereby the men and women who had laboured under a 'domestic' putting-out system of woollen production were gradually drawn into an 'industrial' factory system (as masters, as hands or as a proletarianised labour reserve). This is a compelling story, full of drama, politics, hardships, and Gregory admits that it can be read as 'a conventional narrative' (*ibid.* p. 15) because he tries not to obscure 'empirical materials' behind 'the formal system of concepts which make their arguments possible' (cf. what occurs in Dear and Moos, 1986; Moos and Dear, 1986).

Nonetheless, Gregory does want to insist that 'the problematic of structuration . . . informs the architecture of the text' (1982b, p. 15), and to be more specific he suggests that it does so through three 'decodings' of what 'regional transformation' entailed:

1. It was a '*continuous* movement effected through the *bounded* activities of *reflexive* human subjects': in other words, the gradual shift in the workings of the woollen industry was not effected overnight, but was 'made' through 'the day-to-day flow of practical life – working at the loom or carousing in the tavern, riding out to the cloth hall or walking back from the mill' (*ibid*. p. 18).

2. It was transition whereby 'labour is subordinated to capital', and during which 'a series of changes in the *labour process* meshed with other sequences turning in *politics* and in *ideology*. The precise alignments between them were not determined by the rotations of the economy but by the outcomes of *situated social practices*' (*ibid*. p. 21): in this latter respect (2) connects up with (1).

3. It was a transition crucially bound up with 'the changing time–space intersections of social practices', such that 'the transformation of the woollen industry of the West Riding of Yorkshire is to be understood as the constant repetition of characteristic *time–space routines* – the intertwining of paths and projects – through which the structures of the domestic and factory systems were fleetingly engaged and regularly reconstituted' (*ibid*. p. 23): and here (3) connects up to (1) and (2).

These three 'decodings' progressively introduce into the account of 'regional transformation'

1. thinking and acting human agents leading their everyday lives of work and leisure;

2. situated social practices through which specific intersections between economy, politics and ideology are worked out; and

3. a time–space geography of these practices that was deeply implicated in the specifics of how economy, politics and ideology (and hence the whole package of changes involved as the domestic system shifted into the factory system) were co-ordinated.

And here Gregory is clearly conjoining structurationist and time-geographic arguments in a manner akin to that already explained.

There are various ways of further 'unpacking' Gregory's narrative, but perhaps we can identify a series of spatial scales at which he identifies geographies of regional transformation:

1. The *local* scale: and here he traces the daily and yearly lives of an individual out-working weaver such as Cornelius Ashworth (his time at the loom, his time using scribbling- and fulling-mills, his time visiting the Halifax cloth hall, his time on hay-making – and in so doing illustrates how these activities all drew upon the resource of space as well as that of time). And Gregory uses this information to present a time–space diagram similar to that produced by Pred for his Boston merchant (see Figure 4.11).

2. The *regional* scale: and here he traces the geographical patterns – and their changes through time (chiefly late eighteenth and early nineteenth century) – in numbers of out-workers, numbers of clothiers, numbers of the various different types of mill (notably comparing water- and steam-powered mills), numbers of factories, ratios of labour to (machine) power, and so on within the Yorkshire woollen district. In addition, Gregory traces patterns in where out-workers and clothiers went to trade their wool (and hence he

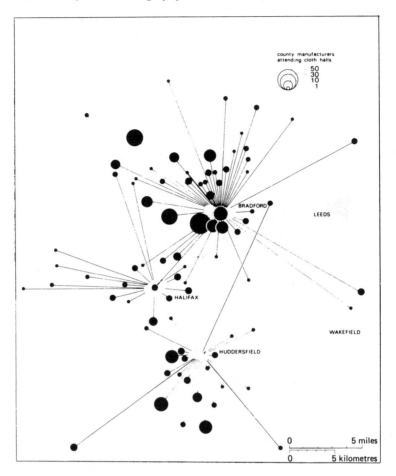

Figure 4.12   Cloth-making systems: Bradford, Halifax and Huddersfield c.
1822 (Gregory, 1982b, p. 118)

ascertains the intra-regional geography of trade/business transactions).
Figure 4.12 provides an example of these regional movements: in this case,
movements to centralised cloth halls.
3. The *national* scale: and here he traces the 'performance' of the Yorkshire
   woollen district in relation to the contemporaneous performance of other
   cloth-making districts in the British Isles and beyond (see the relative employment changes shown in Figure 4.13, for example). Here too Gregory
   seeks to establish trade flows linking the Yorkshire woollen district to areas
   from which wool was sometimes imported and to which finished cloths
   were then exported (flows with a significant international component).

We should also note that, whilst the above account focuses principally on
economic geographies, up to a point Gregory attempts to tease out corresponding (though not always neatly overlapping) *political* and *ideological*
geographies to go alongside the economic.

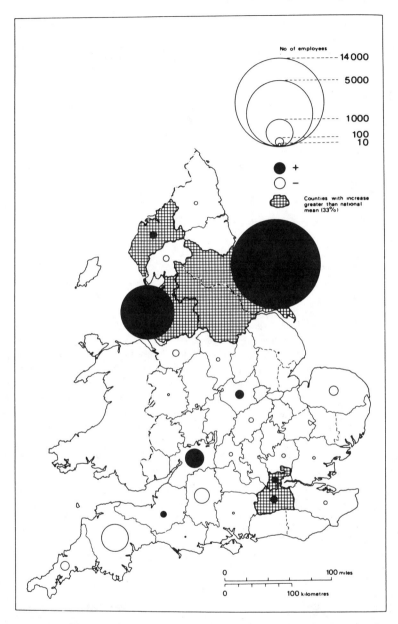

Figure 4.13 Change in employment in woollen mills, 1838–50 (Gregory, 1982b, p. 62)

For Gregory, though, the important thing is not so much the geographies of these three spatial scales seen in isolation, as the ways in which the time–space rhythms of one 'meshed' with the time–space rhythms of the others: when (say) the time–space 'inefficiencies' in the domestic system – the effort expended as out-workers tramped between their cottages, the mills and the cloth halls –

placed limits upon the cloth production of Yorkshire relative to production elsewhere; or when (say) recession in the national 'space-economy' forced a 'rationalisation' of the Yorkshire woollen industry, leading to the replacement of labour by machinery in specific localities favourable to water or steam power, or to the elimination in some localities of the time–space 'inefficiencies' in the domestic system by entrepreneurs bringing together different stages in the cloth-production process under one (factory) roof. And, again, we should indicate that at the same time as recovering these more *economic* dimensions to how the geographies of different spatial scales connected up one to another, Gregory also recovers more *political* and *ideological* dimensions.

Throughout these inquiries Gregory claims to be illuminating the continuous 'structuration' of the Yorkshire woollen industry as this occurred through the complex interconnectivities of agency and structure as these were 'played out' both through time and across space, and in the process he arguably takes structuration beyond Giddens's own conception – in part simply by putting some empirical 'flesh' on the theoretical 'bones' – and thereby presents us with some key critical analyses of Giddens's work. It is to the more evident lines of Gregory's critique we turn now.

Perhaps Gregory's overarching critical concern with Giddens's work is that social life is made to appear 'more coherent' than it really is because the full implications of incorporating time–space theory into structuration have not been properly realised (see also Chapter 6). So far as time is concerned, Gregory criticises Giddens for treating time-geography merely as a 'series of templates' for delimiting pathways in time and space, whereas Hägerstrand's work is more ecological and deals too with the 'deeper resonances' between time-geography and phenomenology: in Heidegger's (1971) terms, 'the making room for things in ways which allow them to be the things that they are' (see also Chapter 3). The time–space path intersections of Giddens's structuration, then, do not fully reflect the 'fundamental bonds between life, time and space' spoken of by Hägerstrand (1974), who argues that particular projects of action have to be accommodated within the pressures and opportunities experienced by the individual within bounded times and regions. Gregory concludes that it is therefore essential to identify the 'time–space modalities' that 'enable projects to hook up in particular configurations' and are the vital conceptual bridge between the spatial and the social. This, he argues, make time-geography 'a much more inclusive model than Giddens allows' (Gregory, 1989b, p. 196). These reservations lead on to a series of related criticisms.

For instance, Gregory (1989b) offers four 'glosses' on Giddens's theorisation of *time–space routinisation and reproduction*:

1. Giddens's recognition of the importance of routine social interaction tends to under-emphasise the importance of *strategic intentionality* (see, for example, Scott and Storper, 1986) and *discursive consciousness* (see Storper, 1985) in understanding action and agency. Strategic intentionality can be illustrated by the work of Massey (1984), who has linked technical changes in production to the corporate strategies of industrial capital to show how spatial divisions of labour are embedded in successive phases of investment. An illustration of discursive consciousness is provided by Lash and Urry (1987), who suggest that members of the service class have achieved a range of class privileges (including that of being able to colonise certain residential

spaces) by the appropriation of specific forms of discursive knowledge (educational, professional, expert and so on) to themselves. In both cases, a more complex account of the changing geographies of 'strategy' and 'discourse' that underpin routinisation and reproduction seems to be required.

2. Giddens's version of routinised action and interaction does not show how the separations between action that is *taken-for-granted* and that which is *accepted-as-legitimate* can be made wider in the course of routines. As Gregory (1989b, pp. 200–1) states:

> Just because actors are bound in to time-space routines, it does not automatically follow that they internalize the values of the social institutions which are reproduced through them. . . . [T]o accentuate the recursiveness of time-space routines without opening a conceptual space for de-routinization 'from the inside' is, I think, to short-change the transformation model embedded within structuration theory.

Thus, routines can become progressively disengaged from the continual rounds of social reproduction: otherwise social change would only occur in a 'discontinuous process of sudden and spectacular transformation'.

3. The dislocation of routine can occur through sudden critical situations of crisis. Habermas (1976) stresses that such crises are both 'caused' and 'experienced', but Giddens focuses almost exclusively on the experiencing of crisis rather than on the cause. Both sides to the understanding of crises are required if we are to realise the existence for 'multiple and competing possibilities for change', which occur within that crisis. In other words, structuration would benefit from an account of the *conditionality* of crisis.

4. Giddens's suggestion that routine and social reproduction occurs in locales tends to obscure the notion and importance of 'place', a notion that suggests a depth of meaning, belonging and interpretation that the use of 'locale' cannot convey (see Chapter 3). Pred's work on the becoming of place in a structurationist framework achieves some, but only some, linkage between the two concepts.

Gregory's critique also ranges over Giddens's theorisation of the relations between *time–space distanciation and social reproduction*. Here he raises two fundamental objections. First, Giddens's concept of distanciation tends to regard space merely as a 'gap to be overcome'. He thereby takes little or no account of both the *production of space*, that is the production of *material spatial structures* through which the agencies allowing space to be 'overcome' (writing, money, transport networks) are put into place, and the *representation of space* through which spatial organisation and the 'overcoming' of space are given symbolic meaning. Second, time–space distanciation is seen in terms of the way in which power is generated through the extension of the structures of domination (see Figure 4.3). Thereby it is argued that Giddens undervalues signification and legitimation structures that elsewhere he treats as separate and equally important components of structuration. These objections pave the way for three specific criticisms of the structuration thesis:

1. Distanciation runs the risk of 'over-totalising' social life (see also Chapter 6). Gregory suggest that Giddens's notion of gradually widening systems of social interaction through distanciation obscures the complexity and volatility of extending time and space away from face-to-face routine

encounters. He illustrates this criticism by reference to the landscapes of contemporary capitalism that are

> riven by a deep-seated tension between polarization in place and dispersal over space. On the one side, constellations of productive activity are pulled into a structured coherence at local and regional scales, while on the other side these same territorial complexes are dissolved away through the restructuring and re-synthesis of labour processes. The balance between them – the geography of capital accumulation – is drawn through time–space differentiation as a *discontinuous* process of the production of space.
>
> (Gregory, 1989b, p. 207)

Giddens then displays little fascination with location, and may indeed be guilty of tacitly accepting a dualism between location and interaction. He is interested in locales, particularly those dominant arenas of interaction he calls 'power containers' where authoritative and allocative resources are kept. But his fascination is more with cities (as the important locales for class-divided societies) and nation-states (as the important locales for modern societies) than he is with identifying a *hierarchy* of locales in these and other societies (and this remark links back to our previous discussion of the distinction between locales as containers of interaction, and locales as locations). His notion of power may therefore be too tightly contained in his stress on the city and the nation-state.

2. Gregory also suggests that Giddens emphasises the changing conceptions of time more than those of space (see also Soja, 1989). Yet spatial concepts are of equal importance to those of time in the constitution of society. This may be illustrated by the way in which the development and interpretation of printed maps sustained the political and economic visions encapsulated therein, to consolidate the nation-state and thereby to pattern the system of the modern world. By theorising distanciation with an emphasis on time rather than on space, Giddens risks losing sight of some of the modalities through which power gains entrance to social constitution.

3. Giddens is criticised for not permitting his notion of distanciation to account for the complexity of collective resistance in different societies:

> Giddens has so far not shown how the transformative capacities of human beings vary according to the specific circumstances in which they find themselves and through which they are constituted as knowedgeable and capable human subjects. His preoccupation with an abstract account of human subjectivity . . . prevents him from following through the consequences of his own insight into the way in which social structures enable as well as constrain.
>
> (Gregory, 1989b, p. 213)

Gregory's reservations about structuration culminate in a discussion of structuration and critical theories (see Figure 4.14) where he sketches out a picture of Giddens's present out-working of the structuration of modern society. The modern world is seen as being shaped by the intersection of capitalism, industrialism and the nation-state, and the conjunction of these three components spreads out over four axes. Starting from the centre of the diagram, there are four levels of concern:

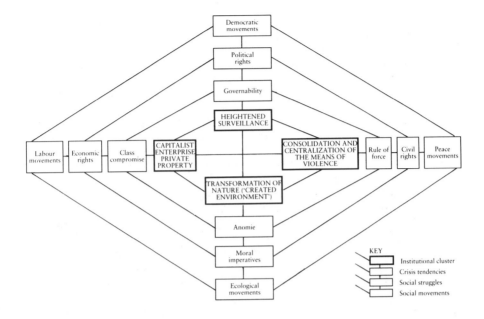

Figure 4.14 The grid of modernity (after Giddens) (Gregory, 1989c, p. 374)

1. The four *institutional clusters*, which supposedly outline the structures of modernity.
2. These then inter-relate to typical *crisis tendencies* of which the top-left quandrant refers particularly to capitalist societies, the top right to socialist societies and the lower half more generally to the crises of industrialism.
3. Crisis tendencies in turn sponsor particular *social struggles*.
4. Social struggles are characterised by specific *social movements*.

It is Giddens's contention that all social movements and spatialities of social life can be located within this schema (Giddens, 1985a), but – and as Gregory (1989c, p. 375) points out – 'some parts of this map are more throughly surveyed than others'. The geographies of power, territoriality, the nation-state and the modern world system are well represented, but that of the uneven development of capitalism and of crisis tendencies are undervalued. In particular, there is no in-depth account of 'the chronic tension between the immobility of spatial structures and their capacity to stretch across even wider spans of time and space' (*ibid.*). Perhaps most important of all so far as the geographical debate on the 'modern' and the 'postmodern' is concerned, Giddens's insistence that distanciation is given form and substance by the resources connected with the structures of domination leads to an undervaluing of signification and legitimation structures that were originally held to be co-partners in the structurationist thesis. The relative weighting given to these components of Giddens's theory hence prevents matters of culture from taking a more central role in the understanding of how human agency is transformed. These issues are further discussed in both Chapters 3 and 6, notably where we consider the impact of the various strands of postmodernism on human geography.

A concluding summary of critique

We began this chapter by asking whether structuration theory should be regarded as a mere 'fashion or fad', and whether it is possible to sustain 'middle ground' theorising in the face of inevitable criticism from the disciples of the positions on either side of the dualism that is allegedly being transcended by that theory. The answer to the first question must be that structuration theory has now had a decade of service to social science and currently does show some signs of fading away to make space for the next fashions (perhaps realism and postmodernism). Having said that, the interactive debates between proponents of structuration and (say) postmodernists still seem to have the potential for fruitfulness rather than destruction. The answer to the second question must also be at least partially favourable to structuration theory. Largely due to the prolific efforts of Giddens to progress through his programme of theorisation, responding along the way to the substantive criticisms levelled at it, structuration theory has been sustained as an important force in wider social-science theorising. There have, however, arisen a number of fundamental objections to structuration briefly mentioned here in concluding the chapter (we have of course already identified a number of specific criticisms in the chapter). This list is neither complete nor mutually exclusive, but it does present us with important pointers as to how structuration theory could and should be used in the practice of human geography:

1. Despite Giddens's own claim that in forwarding structuration theory he is 'firing intellectual salvos' rather than exploding a 'big theoretical bang', his own writings, and style of response to criticism, do suggest that what we have here is something approaching 'grand theory'. As Gregory (1990b) notes, 'his proposals are still grand in the sense . . . that they reclaim the grand question for socio-historical inquiry'. Although grand theories are not the same as grand questions, there do seem to be strong interconnectivities between the two.
2. Linked to (1), Giddens's grand questions – and his associated conceptual stretching of social reproduction over time and space – appears too clearcut, and is therefore likely to obscure some of the highly differentiated forms of structure and agency in which human geographers will be interested. As Mann (1986, p. 4) suggests, 'societies are much messier than our theories of them', yet Giddens perhaps has a tendency towards overcategorisation into neat little boxes of understanding. The discussion above of 'power containers' is a case in point.
3. Although regionalisation is an intergral component of Giddens's structuration theory, a major trajectory of critical comment is that the spatial nature of regionalisation is one of the least well-worked-through areas of structuration writing. Consider, for example, the subtle variation between Gregory's (1986) portrayal of time–space relations in Figure 4.3, and his equivalent diagram in Figure 4.15, published in 1989. In the former it is regionalisation that holds the core position, interlinking with societal integration, routinisation, distanciation and contextuality. In the latter, it is the

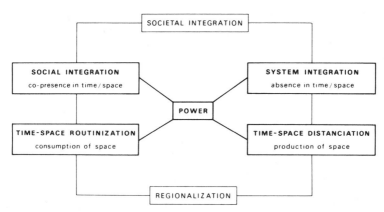

Figure 4.15   The time–space constitution of social life (Gregory, 1989b, p. 188)

notion of power that has been placed in the core of the diagram, and regionalisation replaces contextuality at the foot. This seemingly minor, but actually very important, change in the representation of time–space relations by one of the principal translators of structuration for geographers reflects the direction of Giddens's own thinking during Phase II of his project. If it be an appropriate criticism that Giddens has focused less than adequately on space, then it is geography, rather than Giddens, who has lost out the most.

4. Linked with (3), regardless of Giddens's own protestations on the point, criticism continues to be directed at an apparent polarity of scale in his work. Certainly, the examples used by Giddens in *The Constitution of Society* (1984), for instance, do seem to reflect either the micro-world of day-to-day and face-to-face interaction or the world system at large. And elsewhere the city and the nation-state appear to be essential to the definition of regionalisation. 'Meso-level' *socio-spatial scales* of analysis appear far less commonly. Equally, it has been suggested that structuration theory would be improved by the development of 'meso-level' *concepts*, which could link the 'more abstract propositions about time and space to more detailed investigations of the specificities of history and geography' (Gregory, 1989b, p. 214). Gregson (1989; see also Gregson, 1987a) notes that these 'meso-level' propositions would allow structuration to encompass structures such as patriarchy, racism and capitalism within the overall theorisation of structuration. It might also allow a more hierarchical appreciation of locales.

5. This criticism of the lack of meso-level concepts invites more specific discontent about particular aspects of social reproduction Giddens has seemingly thus far ignored. The glaring example is the omission of any detailed mention of *women and gender relations* in a supposedly wideranging social theory, let alone it paying any regard to particular spheres of social activity usually associated with women. As Murgatroyd (1989, p. 147) argues powerfully, 'not only is half of society (and the relationship between that half and the other half) virtually omitted from his analysis, but also, partly by association half of the activities carried on in the society'.

These comments could be applied almost as well to factors of ethnicity. Specifically, then, Giddens's structuration does not deal directly with the gender division of labour and more broadly (and fundamentally) it does not cover the crucial areas of social life relating to gender and ethnic relations: a criticism regarded by Murgatroyd and many others as 'a major flaw in his work to date'.

6. Giddens's treatment of the concept of *environment* has also been subject to critical scrutiny. It can be suggested that his identification of a 'created' (that is, a planned or built) environment, representing an important cluster of institutions, is too neat to describe the chaotic and often confusing intersections between capitalism, urbanism and spatial territoriality. More-over, the links Giddens makes between the created environment and the 'natural' environment appear a little deterministic (but perhaps not quite functionalist) in the routines they are said to engender. Thus, Saunders (1989, p. 232) detects in Giddens 'a rather conservative romanticism which assumes without proof or argument that a routinized existence in a "natu-ral" environment (whatever that is) generates harmony and "symbiosis", while a routinized existence in a humanly transformed or created environ-ment generates anomie and disrupts ontological security'.

7. Crucially, in assessing the relevance of structuration to the practice of hu-man geography, the ontological trajectory of the theory means that the questions it raises and the categories it proposes are concerned with validat-ing ontological claims rather than genuinely informing empirical inquiry. Gregson (1987a; 1989) stresses that on these grounds it is unreasonable to expect that structuration theory should offer us instantly operable empirical research questions or analytical categories. Indeed, she believes that 'to transfer structurationist concepts directly into empirical analysis is miscon-ceived' (Gregson, 1989, p. 246). This relationship between structuration theory and empirical inquiry can be looked at in two ways. Either it is an inappropriate criticism to say that structuration does not transfer directly across to the empirical research of human geographies, because such a transfer is beyond its terms of reference. Or it might be said that structura-tion ought to have been developed in such a way so as to achieve more immediate relevance for empirical work. This latter view relates to the suggestions in (4) and (5) about the introduction of meso-level concepts. After all, if structuration theory is unable to gain purchase on the structures that exist in the contemporary world, then of what use is it?

8. A final criticism that many interpreters of Giddens have perhaps been too charitable – or perhaps too personally 'implicated' in – to articulate con-cerns the way in which a particular 'discursive community' has grown up around structurationist thinking. Despite our best endeavours to the con-trary, it will be evident from this chapter that structuration is characterised by a complex language: a language that has been continually nourished by a discursive community proposing and discussing concepts with its own metaphors and rhetorics. Although it is not necessarily any better or worse than the languages of the communities surrounding (say) realism or postmodernism, the vocabularies of structuration do sometimes seem cap-able of obscuring rather than elucidating the concepts concerned and the substantive issues 'beyond'. Some will find working with these vocabularies trouble-free and even appealing. Gregory, for example, talks at one point

about the baroque intricacy of Giddens's conceptual grids: a comment that indicates both an ability to obscure and an ability to impress. Generally, though, Giddens's discursive language of structuration appears devoid of any stylistic features that might give his work a more cultural interest and character. In this way he may do a slight disservice to his students but not to his peers: but as he has so few peers, this may still be a telling criticism.

Giddens (1989b) offers a response to most of these points, but in the final reckoning it will be for each individual to assess the influence that Giddens's structuration will have on their work. Giddens has provided an important point of dialogue between concerns over structure and those over agency, and has acted – or at least could have acted – to introduce doubt into particular philosophical positions. Overall, it may be that the most important role for structuration is to present us with a series of warnings about how we should *not* approach the complex matters that constitute human geography (that is, by avoiding the excesses of either overly structure-orientated or overly agency-orientated approaches), as opposed to presenting us with a comprehensive blueprint for how we should carry out our studies.

5

REALIST APPROACHES IN HUMAN GEOGRAPHY

All roads lead to realism?

Is it really true, there are elephants, lions too, in Picadilly Circus?

(Jethro Tull, 1971)

I suspect that we are all 'realists' now, with about as much understanding and sophistication as we were all 'Marxists' in the early 1970's.

(Archer, 1987, p. 390)

During the 1980s and early 1990s there has been a serious engagement by human geographers with the philosophy, theory and methodology of *realism*. It is important to note at the outset of this chapter, however, that this seemingly recent attraction to realism has, in fact, much deeper historical roots. Thus since the early history of philosophy, especially in Platonic–Socratic thought, there has been a notion that there exists a 'real' world of physical things, which is independent of our senses and therefore is independent of our perception and cognition of those things. In this form, realist philosophy is in direct opposition to *idealism* (see Chapter 3), which claims that the world is only made known through, and is therefore constituted by, our 'ideas' (and this term refers to senses, perceptions and interpretations). This opposition between early realism and idealism has been neatly captured by Relph (1989) (see Figure 5.1).

The translating of this early realist philosophy into geographical writing has been discussed by Gibson (1981, pp. 152–3), who adopts the term *naïve realism* to describe the outcome:

For the Naive Realist, or common-sense geographer, the mind grasps the world in a simple effortless process, something we do all the time. True, our perception can occasionally go wrong: we fail to see the 'blue' hue to the distant mountains. Or the distance we walk between the classroom and library 'seems' much shorter than the two hundred yards confined on the campus map. Still these are minor exceptions that we discuss because of our inattention. No doubt, with the collaboration of others, with the aid of improved instruments, we can establish the 'real' nature of the physical world. Geographical facts of observed phenomena and changes within them

Figure 5.1   Relph on realism and idealism

Relph (1989) highlights the work of the seventeenth-century philosopher, John Locke, who differentiated between 'primary' qualities of
things (e.g. solidarity, extension, figure, mobility) and 'secondary'
qualities (e.g. colours, sounds, tastes). For Locke the *real* world is
thereby composed of primary qualities, existing independently of
humans, while another 'insubstantial' world is composed of the secondary qualities delivered by our senses. The real world therefore lies
behind, or above, or below the environments of everyday experiences.
By contrast Relph draws on the writings of the early eighteenth-century
philosopher George Berkeley, who suggested that the reality of things
lies in our perception of them. What we know of things we know only
through our senses and ideas – this was the basis for the school of
thought termed idealism. In this form there is strong opposition between
the two philosophies. As Relph comments, 'for the realist idealism is a
form of lunacy, for the idealist the assumptions of realism are a bizzare
fantasy' (p. 280).

can be objectively established and any questions of unseen entities, problematics,
abstract forms or subjective impressions are irrelevant. With due precaution, so say
common-sense geographers, we can know a place as it really is, as the topography
looks, as the soil feels, as the water tastes and so on. We may understand that the
measures we use in objective studies, like two hundred yards, for example, are
products of our minds and not the external world alone. Yet we also understand that
they are not the products of private illusions either. Each measure is neutral and exists
independent of our thinking, so far as Naive Realists are concerned. Thus, according
to this logic, conveying the world requires nothing more than a description of its facts
and most assuredly it does not require assumptions of 'hidden' entities.

Gibson charts the progress of naïve realism through 'capes and bays' geography, via the resource inventories of, for instance, Dudley Stamp, to a period
of prominence in the geography of the inter-war years. He attributes to it both
the merits of creating a class of professional geographers, and the disadvan-
tages of the demoralisation that creeps in when mechanical inquiry is divorced
from surrounding philosophical debates. Ultimately, he suggests, it was the
excesses of extremist naïve realism that were at least in part responsible for the
scientific revolution of the 1960s.

It is against this background that we should view the resurgence of realism
as a philosophical guide to research in human geography in the early 1980s,
for this more contemporary adoption of realist philosophy has gone well
beyond its naïve predecessor. Realist philosophers such as Roy Bhaskar suggest that, 'yes there is a *real* world', but that some of the most significant
components of this world are not immediately observable. Indeed, there are
experiences, systems and structures (whatever precise definitions we might
lend to these terms) that cannot be discerned in a straightforward measurable
manner, but that not only exist but in some sense also constitute much that
occurs in the human world. And when seen in this light contemporary realist
philosophy might be said to have incorporated notions of idealism – in
recognising that some 'things', whilst being accepted as 'real', are still only

knowable through our concepts of them – to which naïve realism was opposed. It is this crucial change of philosophical direction away from naïvety that underpins the recent surge of interest in realism.

The recent surge in attention given to realist approaches is marked by the extent to which the formative contributions by Roy Bhaskar (1975, 1979, 1986, 1989) on the philosophy of realism, Russel Keat and John Urry (1975) on the theories of realism, and Andrew Sayer (1982, 1984, 1985a, 1985b) on the methodology and practice of realism, have been prodigiously referenced as the sources of both a new way of thinking and a novel way of reinterpreting what is already being practised in human geography. In the sense discerned by Archer (quoted at the beginning of this chapter) realism has had an extraordinarily wide appeal, then, but there is a suspicion that for some its appeal has been founded on a less than full recognition of its complexity. Indeed, as Sarre (1987) has suggested, there are two interpretations of realism that have proved consistently problematic to geographers (and other social scientists): first, the philosophical underpinnings of realism are complicated, and represent quite a different way of looking at the world from that of other positions such as positivism, Marxism and humanism; and second, it is claimed that realism is the *de facto* foundation of much scientific research, meaning that some of its assumptions have already been internalised and consequently that some social scientists do not find it easy to characterise it as a new or distinctive approach to research methods. A potential pitfall inherent in this latter problem is that quasi-realist interpretation can be 'read into' research, which does not take full account of the philosophical complexity of the former. It is therefore important for us to 'unpack' the propositions of realism with care lest we fall into this trap of subscribing to the realist position in name but not in nature.

Aside from its prominent position over the last decade, and the crucial suggestion that it has at times been adopted too lightly and superficially, realism is also important to human geography because of its associations, both potential and actual, with 'structuration theory' (see Chapter 4). As we will see, realism entertains some ontological overlap with structuration theory (the work of Bhaskar spans both, for instance) although its main task has been to begin to offer an epistemology: that is, a theory of the nature of knowledge and a justification of belief. Realism has thus become, rightly or wrongly, part of the search for 'middle ground' approaches to human geography we described at the beginning of Chapter 4. Here it is important to distinguish between the *philosophical* position of realism – its concern for the fundamental 'building blocks' of reality and of how we can acquire knowledge about these blocks – and a more *social-theoretical* position such as that of Marxism, which is concerned to specify much more precisely and concretely what the human world contains (e.g. modes of production, social classes, states) and how its contents operate. Thus realism as philosophy and Marxism as social theory can indeed be seen as compatible, just as it might be claimed that realism can be the underpinning philosophy of structurationist social theory and that realism is equally compatible with emerging studies under the postmodernism banner (see Chapters 4 and 6).

So what is this realism that can seemingly be all things to all people if we are not careful? A commonly quoted definition is provided by Gregory (1986, p. 387): 'Realism is a philosophy of science based on the use of abstraction to identify the (necessary) *causal powers and liabilities* of specific *structures* which are realised under specific (contingent) conditions'. Note here that the contemporary concerns of realism are very different from the earlier versions of naïve realist philosophy. Note also that there are several key terms in this definition that require unpacking (this is a term often used in realism to denote breaking down into more limited and specific meanings for ease of understanding) and we will deal with these terms as we proceed through the chapter. Finally, we need to recognise that the importance to geography of realism in the post-positivist era (see Chapter 1) stems from the critique of the foregoing scientific revolution, and incorporates a rejection of approaches to knowledge that assume quantification *per se* to provide some privileged access to the real world.

Moreover, at their most expansive, proponents of realism claim that all of the post-positivist tendencies in human geography can be interpreted as roads towards realism, be they labelled Marxist, humanist or even other 'mainstream' geographies. In a manner that suggests a very privileged role for realism as the keeper of the geographical tablets of truth, Sayer (1985a, p. 161) has claimed that

> all the above post-quantitative trends are moves towards a realist approach. The changes are clearest in radical and Marxist research in geography. . . . However, this association of realism and radical geography is not a necessary one . . . and acceptance of realist philosophy does not entail acceptance of a radical theory of society – the latter must be justified by other means. Some aspects of humanistic geography might also be taken to be convergent with realism and occasionally even mainstrean geographers stumble across realist methods, if only temporarily and unknowingly.

Evidently, then, realism is built first and foremost on a rejection of positivism and, as with other philosophical standpoints, realism is perhaps best initially understood in relation to its critique of another philosophy: in this case positivism. We turn first, therefore, to the work of Bhaskar and others who present an extended critique of positivism and propose a realist philosophy.

A realist philosophy

'Roy Bhaskar is one of the most original and versatile British philosophers of the post-war generation, and one of the very few whose work is likely to have a really lasting and world-wide impact on philosophy and social science' (Outhwaite, 1989). It has been Bhaskar who has led the way in contrasting the empiricist approaches of positivism with the emergent philosophy of realism. At the outset Bhaskar suggests that realism should be a mix of

1. non-atomistic ontology; and
2. non-empiricist epistemology.

This suggests, in the first place, that realism represents a 'complex ontology' (Outhwaite, 1987), in that it allows for the existence of structures, processes and mechanisms that can be revealed at different levels of reality and that mean that reality is not conceived of as closed simply around observable events and phenomena. Scientific observation may not always be correct: atoms, for example, may at some stage be found not to be a-tomic in nature

Figure 5.2   Four misconceptions about knowledge (Sayer, 1984, p.17)

1. That knowledge is gained purely through contemplation or observation of the world.
2. That what we know can be reduced to what we can say.
3. That knowledge can be safely regarded as a *thing* or *product* that
   can be evaluated independently of any consideration of its production and use in social activity.
4. That science can simply be assumed to be the highest form of
   knowledge and that other types are dispensable or displaceable by
   science.

but rather themselves made up of complex structures. Realism thus allows
for the reformulation of what is 'real' as different properties of a situation,
or a 'thing', come to light. The non-empiricist epistemology permits realism
to challenge the empiricist conceptions of causality, which have become
ensconced in many other approaches to human geography.

Andrew Sayer (1984) provides a good illustration of non-empiricist epistemology in suggesting four basic misconceptions about knowledge that
should be challenged by realism (see Figure 5.2). Each of these assumptions he
claims can be answered. First, knowledge can come from *participation* as well
as from observation, through working for change, interaction with others
involved with common resources and language (see Habermas, 1972). Second,
spoken and written forms of language are not the only ways to communicate,
appreciate and apply knowledge. The everyday practical skills of knowing
(Giddens's routine 'reflexive monitoring'; see Chapter 4) – for example, the
feelings (as bound up with sight, sound, smell and sense) that give you knowledge about being in a large crowd at a sports stadium, or about a view of the
natural environment unfolding before you or about whether a place represents
security or threat – are often not capable of verbal description, yet are as
meaningful to the human subject as speaking or writing (see Chapter 3). Third,
knowledge is not a finished product to be plucked from outside of us and
analysed and stored inside of us. According to Bhaskar (1979), knowledge is
both an 'ever-present condition' and a 'continually reproduced outcome' of
human agency. Fourth, science cannot be assumed as the highest level of
knowledge because science has not readily coped with participation, feelings
and continually reproduced knowledge as proposed in the challenges to the
first three misconceptions.

The positivist approaches resting on these misconceptions posit that if we
are to explain processes we must discover the regularities that govern the
relationships existing between observable events and phenomena. Realism on
the other hand views causality in terms of the 'nature' intrinsic to what is being
studied, the interactions between that and other things, and the causal powers
and liabilities involved. The realist argument contends that if we wish to
explain why certain things behave in a certain manner, then we must understand both their internal structure and the mechanisms and properties that
enable them to produce or undergo particular changes when placed in contexts
where they interact with other things.

Figure 5.3 The three domains of reality (Bhaskar, 1975, p. 56)

	Domain of real	Domain of actual	Domain of empirical
Mechanisms	✓		
Events	✓	✓	
Experiences	✓	✓	✓

These general criticisms of positivistic assumptions about the way the world works, and about how we understand causality, are developed by Bhaskar into a claim that positivistic approaches oversimplify and conflate (to use Bhaskar's terminology; see Figure 5.3) the three separate domains of

1. the *real* (mechanisms, which are unobservable);
2. the *actual* (events, which are observable phenomena); and
3. the *empirical* (experiences, of events).

Bhaskar (1975, pp. 56–7) argues his case thus:

> I have argued that the causal structures and generative mechanisms of nature must exist and act independently of the conditions that allow men [*sic*] access to them, so that they must be assumed to be structured and intransitive, ie. relatively independent of the patterns of events and the actions of men [*sic*] alike. Similarly, I have argued that events must occur independently of the experiences in which they are apprehended. Structures and mechanisms then are real and distinct from the patterns of events that they generate; just as events are real and distinct from the experiences in which they are apprehended. Mechanisms, events and experiences thus constitute three overlapping domains of reality, viz. the domains of the *real*, the *actual* and the *empirical*. . . . The crux of my objection to empirical [or naïve] realism should now be clear. By constituting an ontology based on the category of experience, as expressed in the concept of the empirical world and mediated by the ideas of the actuality of the causal laws and the ubiquity of constant conjunctions, three domains of reality are collapsed into one.

Bhaskar's case against positivistic approaches

From the above statement it becomes clear that

1. causality (the real) does not necessarily follow pattern (the actual);
2. causality (the real) does not necessarily follow experience (the empirical); and
3. positivism (Bhaskar's empirical realist category) collapses the real, the actual and the empirical.

However, critics of realism would wish to pose the question of how Bhaskar knows that his three-tiered version of how reality works is 'true', since his only access to reality – just like everybody else's – is only through his own apprehension of it. Here his own 'experiencing' of real observable things (as connected up to his own theorisation of them) is arguably not problematised enough, and it might be objected that his realism simply reproduces the arrogant claims to 'know the world as it really is' that have bedevilled positivist philosophy.

In order to overcome these difficulties in taking full account of each of the three domains of reality, Bhaskar develops the philosophy of *transcendental realism* (sometimes also known as *theoretical realism*). Here, the objects of knowledge are the structures and mechanisms that generate phenomena. These objects of knowledge are not observable events and phenomena (as viewed in empiricism); they are not the human constructs imposed on phenomena (as viewed in idealism); but they are supposed to be real structures and mechanisms that exist independently of our knowledge and experience and of the circumstances that permit us access to them. It is therefore the distinctions between the domain of the real, the actual and the empirical, and the interactions between them, that lie at the core of transcendental realism, and in turn form the basis of the realist critique of empiricist laws of actuality through regularity.

Russell Keat and John Urry (1982) add to the power of the realist argument by comparing realism with the other common refutation of the empiricist approach: that of *conventionalism* (stressing that knowledge is nothing but 'conventions' of what given sets of scholars deem to be 'true'). Conventionalists such as T. S. Kuhn also argue that the truth or falsehood of a particular theory cannot be fully determined by empiricial data, and they go on to suggest that all observations are influenced by the theoretical and ideological leanings of the observer, and that the acceptance or rejection of theories will thus be susceptible to moral, aesthetic or political prejudice. Keat and Urry emphasise, however, that as a sophisticated (transcendental or theoretical) realism does not grant credence to the dominance of empirical testing of theory, the realist position is relatively undamaged by – and can in a sense take on board (see below) – the conventionalist critique. In fact, this claim by Keat and Urry merely plays out a little further the complexities of the realism–idealism dispute that originated in Berkeley's rejection of Locke's philosophy in the eighteenth century (see Relph, 1989).

Thus far, then, we can recognise (following Outhwaite, 1987), five main principles of a realist philosophy. The first three are ontological:

1. There is a distinction to be made between *transitive* objects of science (models, concepts, etc.) and *intransitive* objects, which are the real entities and relations that comprise the natural environment and the social world.
2. Reality may further be stratified into the domains of the *real*, the *actual* and the *empirical*. The empirical domain is contingent on its relations with the domains of the actual and the real.
3. Causality is a tendency grounded in the interactions between generative mechanisms in the domain of the real. These interactions may or may not lead to events in the actual domain, which may or may not be observed in the empirical domain.

There follow two epistemological principles:

4. Empiricism and conventionalism are rejected, with emphasis being placed rather with 'real definitions' of the basic properties and nature of a particular entity or structure.
5. The concept of explanation reflects (3) above, and involves the supposition of explanatory mechanisms and the related attempts to demonstrate their existence.

Applying realist philosophy to social science

The debate over naturalism

So far we have summarised the basic philosophical principles Roy Bhaskar and other have derived from a critique of both empiricism and a positivist search for the real. These principles have been couched in very generalised terms, and may as yet seem difficult to grasp hold of with reference to the substance of the study of human geography. Things become a little clearer, however, as we move to the next stage of the realist journey when these broad philosophical principles begin to be applied more specifically to social science.

There has been a vigorous debate among key realist thinkers over how to translate realist philosophy into social science terms, and the core question in this debate has been that of *naturalism*. Previous positivist claims that all science (whether focused on natural or social worlds) can proceed according to the one model of conventional natural science, thus bringing about a unification of the law-seeking ambitions and attendant languages of the 'scientific method', are rejected. By many realist philosophers still hold to a form of naturalism that claims it is possible to conceive of basic similarities between the concepts and practices of social and natural sciences. Indeed, Bhaskar (1979, p. 3) suggests that 'it is possible to give an account of science under which the proper and more-or-less specific methods of both the natural sciences and social sciences can fall', although this support of naturalism (*ibid.*) 'does not deny that there are significant differences in these methods grounded in real differences in their subject matters and in the relationships in which their sciences stand to them'. Bhaskar suggests that there are properties in society that are legitimate objects of knowledge from a realist perspective. Societies, he argues, cannot be reduced to the collective actions of individual people. Rather, there exist social forms that are a fundamental or *necessary* condition for any intentional action. Social forms exist before we have knowledge of them and therefore have autonomy as objects of knowledge and investigation. Social forms also have causal power, in that they are capable of influencing certain events and experiences, and hence they are present in the domain of the real. Social forms can, however, be influenced by the events and experiences of human agency, so Bhaskar recognises a 'transformational model' of social reality, which sees society both as a causal condition and as being continually reproduced and occasionally transformed by human agency. There are, of course, clear parallels between this model and Giddens's notion of the duality of structure and agency that is fundamental to structuration theory (see Chapter 4). Bhaskar therefore claims that both social science and natural science can root themselves in similar foundational ontological and epistemological assumptions (assumptions about the basic properties of 'real things' and about how we can acquire knowledge about these 'real things'), but he warns that these assumptions must be thought through afresh and not simply ransacked from existing empiricist and positivist philosophies of natural science.

This hope of covering both social science and natural science under one set of ontological and epistemological assumptions is disputed by Rom Harré (1986), who presents an *anti-naturalist* conception of 'social being' from a realist viewpoint. Harré (*ibid.* p. 237) suggests that there exist certain social structures that are commonly perceived and acted upon by human agents, and that an anti-naturalistic epistemology is required for an understanding of structure and agency in these terms:

> The fact that both natural and social sciences use models in the same way may suggest misleadingly that they share a common epistemology. The differences emerge when we compare the relations of fact to theory in each kind of science. In the social sciences facts, at the level at which we experience them, are wholly the creation of theorizing, of interpreting. Realists in social science hold, and I would share their belief, that there are global patterns in the behaviour of men [*sic*] in groups, though as I have argued we have no adequate inductive method of finding them out.

Harré's concept seems close to idealism (see above and Chapter 3), and emphasises the role of structures as a series of imperatives and constraints rather than as a set of inter-relationships between structure and agency that are mutually reproductive and potentially transformative. From a common foundation of realist philosophy, then, it is possible to reach quite different applications to the understanding of structure and agency.

An important contributor to the debate on naturalist and anti-naturalist approaches, and the associated conceptions of structure and agency, has been Ted Benton (1977, 1981). His work seeks to present a critical engagement with Bhaskar's naturalist position, and in particular he focuses on the suggestion that there are three 'ontological limitations' to a naturalist approach, namely

1. social structures (unlike structures in the natural world) are not independent of the actions they govern;
2. social structures (unlike structures in the natural world) are not independent of the conceptions individual agents have of what their actions entail and mean; and
3. social structures (unlike structures in the natural world) are time-dependent, such that the actions they govern are therefore not universal in time or in space.

According to Benton, these principles cause Bhaskar to depart too far from a naturalist position, and so he strives to adapt Bhaskar's ontological limitations so as to make them more compatible with naturalism. The first and third principles require only tinkering changes, but the second, in Benton's view, requires considerable clarification. For example, although agents almost always have some conception of what their action entails, this conception does not need to be correct for that action to succeed. On a wider front, actors in the modern capitalist world are often not conscious of the inter-relations they have with the capitalist structures that can govern their actions. Some structures can be powerful because they are imagined to be powerful, while others are powerful but often unrecognised. Therefore, there are occasions when social structures might be viewed as independent of the conceptions actors have of them, although Benton would agree that such independence is not the norm.

These varying views on the inter-relations between structure and agency in social science are important because they can lead to subdivisions of the realist approach, as different applications are made from common philosophical propositions. Underlying these differences, however, is the degree to which realism represents 'common-sense social knowledge' rather than 'social science'. One of the clearest writers on this point is William Outhwaite (1983, 1987), who comes down firmly on the side of common sense:

> what seems to emerge fairly clearly is that the social sciences remain closer to commonsense thinking, which is anyway more pervasive and powerful in the social world. By this I mean that we have intuitions about the structure of almost all the social processes we may care to think about; these may be right or wrong, but they at least give us an entrée into the subject matter. In physical reality, by contrast, we have intuitions only about a restricted range of phenomena. . . . In crude terms, the social sciences begin with a head-start over the natural sciences, but instead of running straight ahead in pursuit of new knowledge they move around small circles and spend a lot of time inspecting the starting block.
>
> (1987, p. 55)

It is interesting that Outhwaite is here echoing the claims of the eighteenth-century Neapolitan philosopher, Giambattista Vico, who argued that human beings have much more of a chance in understanding themselves and the world they have 'made' than they have in understanding a natural world 'made' by God (for a discussion in the geographical literature, see Mills, 1982). Outhwaite thus regards realism as an 'ontology of common sense', and it follows that common-sense descriptions of social phenomena are a reasonable point of entry for theorising in social science. The fact that our knowledge of social structures is dependent upon our *everyday concepts* of them, and upon the *everyday activities* they produce, hence need not be an obstacle to our understanding of them. Indeed, from this perspective it is entirely proper for us to ask the question of what a particular society must be like in order to produce the activities that occur within it, and the concepts which individual actors have of those activities.

In summary, then, the debates between realist thinkers over naturalist and anti-naturalist approaches to the social world have resulted in a multi-step realist strategy for social science, which may briefly be described thus:

1. Attention is directed to an object of knowledge that will already be defined and described in the language of the day-to-day world (but note that Bhaskar's transcendental approach reveals a desire among some realists to transcend the specifics of particular times and places).
2. This object is 'unpacked', or subjected to redefiniton, so as to highlight any internal complexity and the importance of external inter-relations and interactive tendencies.
3. Questions of social ontology will influence the redefinition of the object. Some objects will be the product of human interpretation (the approach favoured by Harré), while others will be structured by deeper causes (leading to the more structuralist approach favoured by Bhaskar). Ontological disputes are thus not realist disputes, but rather the selected social ontology is fed into the realist strategy in an appropriate manner.
4. The methods used to investigate the object will depend on how it is conceptualised. Explanation of the phenomenon will therefore be contextually specific (and this ties in with humanistic arguments about 'local knowledge'; see Chapter 3).
5. The effectiveness of theory-choice will depend on how well the phenomenon is explained by the selected theory; whether the theory gives good reason to believe in the phenomenon's existence, and whether any equally good alternatives are available with which to explain the phenomenon.

It can be argued that this strategy becomes more tenuous as we proceed through this list. Many researchers will be happy with the idea of careful selection and unpacking of the objects of study ((1) and (2)), but will perhaps become either confused or ideologically or philosophically biased when it comes to recognising which social ontology is most appropriate for the explanation of that object ((3) and (5)), and will duly slip into a pre-existing automatic gear when it comes to selecting an appropriate methodology (4) not least because existing theory can be said to be as 'effective' as any other (5). The importance of realism, then, is both as a way of looking at things, and as a way in which dialogue between different social ontologies can be preserved.

It should also be noted as we come to the end of this section that when realist theory begins to be applied to social science it demonstrates a strong tendency to swing towards idealism. It appears that many realist arguments inevitably do become idealist as they struggle to come to terms with their own insistence on not closing inquiry around the domain of observable events and phenomena.

A convergence with hermeneutics and critical theory?

It is useful at this juncture to acknowledge that the realist critique of positivism can be seen alongside other major critical movements – *hermeneutics* and *critical theory* (see Chapters 2 and 3) – that seek to undermine positivistic meta-theories of the social world. Although it would be incorrect to suggest that an emergent synthesis has been achieved between these three approaches, it is nevertheless interesting to note the meeting points between realism and its partners in the critique of positivism, because it is through these very interconnectivities that Bhaskar has developed his latest version of realism, which he terms *critical realism*.

It was Habermas (1972) who was a principal mover in demonstrating that hermeneutic issues needed to be granted a formative position in social theory. The term was originally used to denote 'the interpretation of texts' – initially the religious writings of the eighteenth century – but it was gradually stretched to cover more general concepts of linguistic understanding. And during the 1960s Hans-Georg Gadamer (1975) drew on the work of Heidegger to give hermeneutics a universal meaning, because understanding became viewed as the fundamental mode of participation for human beings in the world. The interconnectivity between hermeneutics and realism occurs first at the level of the recognition and unpacking of the objects of social inquiry. The gist of Gadamer's argument, for example, is that social sciences 'avoid' a real understanding by incorporating liguistically sedimented truisms in their analysis of structure in society and by employing methodologically alien forms of understanding without proper hermeneutic reflection. Realism, then, must take full recognition of the importance of textual interpretation in its unpacking of phenomena. Equally, both Gadamer and, albeit in a different way, Habermas demonstrate how hermeneutic problems are more pervasive in the historical-social sciences, and argue that social investigation requires a more nuanced form of study capable of allowing varied possibilities for interpreting historical events. As Giddens (1976, 1979) admits, the need for carefully reflected and potentially varied interpretations of the historical does limit the explanatory power of the historical and exposes the shortcomings of any 'oversimple' revelatory model of social science (referring both to the interpretative and the structural 'polarities' of sociology which he attempts to span within structuration; see Chapter 4). These

hermeneutic principles will also interconnect with realism at the level of understanding contextual specificity when deciding on appropriate methods of investigation for the object of study. The major area of divergence between
hermeneutic principles and realist philosophy is in their application to structure
and agency in society. Social processes cannot be reduced solely to the matter of
linguistic communication, so realists such as Bhaskar would argue: just because
interpretation and understanding are important parts of the human dimensions
to agency, and just because the access of individuals to their own social worlds
cannot but take place through their own interpretative processes, it cannot be
assumed that interpretation and understanding are all that exist in society. Structures in society are produced and reproduced partly through the interpretations
given to them by human agents, but they are also produced and reproduced by
deeper causes, which are not always recognised by these agents. Therefore, parts
of the hermeneutic critique do ally themselves with realist epistemology, but on
this last point hermeneutics follows Harré's anti-naturalistic route rather than
Bhaskar's more structural-naturalistic route.

The critical theory of the Frankfurt School (see Chapter 2) has sought to add
systematic critical procedures to the hermeneutic approach. In particular,
Habermas was compelled by his rejection of technocratic social engineering
and political exploitation to offer a social theory of knowledge that, he argued,
had been suppressed in the 'objectivism' of previous approaches:

> Once epistemology has been flattened out to methodology, it loses sight of the constitu-
> tion of the objects of possible experiences; in the same way, a formal science dissociated
> from transcendental reflection becomes blind to the genesis of rules for the combination of
> symbols. . . . Objectivism deludes the sciences with the image of a self-subsistent world of
> facts structured in a lawlike manner; it thus conceals the *a priori* constitution of these
> facts. It can no longer be effectively overcome from without, from the position of a
> repurified epistemology, but only by a methodology that transcends its own boundaries.
> (Habermas, 1972, pp. 68–9)

Habermas particularly identified this lack of reflection with the trap of positivism, and proceeded to offer a tripartite methodology that would transcend the
boundaries he spoke of:

1. Theories of the empirical-analytical sciences.
2. Hermeneutic inquiry.
3. 'Critical' social science, notably psychoanalysis and the critique of ideology.

This latter 'critical' methodology he sees as being governed by an 'emancipatory
cognitive interest' operating through self-reflection and challenging 'ideologically
frozen relations of dependence' that should, at least in theory, be capable of
transformation. Many commentators see no real conflict between Habermas's
insistence on the input of specific cognitive interests and Bhaskar's transcendental realism. Indeed, there has perhaps been a movement in Habermas's position
away from hermeneutics towards an analytical (realist?) viewpoint, albeit a
movement rejected by many 'critical theorists'. Realism can accept such interests
as a hypothesis but continue to insist on the importance of a domain of intransitive objects as the transcendental basis of social scientific practice.

There are areas of potential disagreement between the critical theory of
Habermas and Bhaskar's realism (see Figure 5.4), but more importantly they
appear to head in the same direction given Bhaskar's (1980, 1982) insistence

Figure 5.4   Areas of potential disagreement between Habermas and Bhaskar (Outhwaite, 1987, pp. 91–2)

1. 'Habermas is certainly committed to rejecting any realist position which retains a correspondence theory of truth. This theory "attempts in vain to break out of the linguistic realm in which alone the validity claims of speech acts can be clarified". Here however there is no apparent disagreement with Bhaskar's position.'
2. 'he would undoubtedly want to reject a realist naturalism which neglected hermeneutic issues. Again, it does not seem to me that this charge can be raised against Bhaskar's version of naturalism.'
3. Habermas would reject any realism that upheld the fact-value distinction in its traditional form. 'Here again, however, there is an important area of covergence with realism around the theme of emancipatory critique.'

that all science must be critical; and that in social science this critical edge should have the potential to become emancipatory. This harnessing of aspects of the work of Habermas and Bhaskar, along with that of others such as Giddens and Bourdieu, provides the energy for a new set of approaches to social science requiring a plurality of methods, a recognition of 'non-scientific' thinking about action and human society, and a limiting of reductionist social theories. It could be that a realist philosophy of science presents a suitable meta-theoretical framework for these initiatives, or it could be that realism is itself paving the way for the fragmentation of postmodernism described in Chapter 6. If realism is seen as breaking down the ontology of the human world – from atomistic event and phenomena-centred ontologies – to a more nuanced recognition of many different but interconnected components to reality, then it could indeed be said to anticipate the 'philosophy of difference' built into postmodernism as attitude (see the second half of Chapter 6). However, if realism incorporates a faith in the researcher's ability to stand outside of history and geography, alongside a belief in the researcher's own ability to develop 'superior' social-scientific theorisations that can slice through the 'misunderstandings' of situated human agents, then it will be heartily rejected by postmodernists.

Realism, social science and human geography

The exacting tasks both of translating realist philosophy into the geographical arena and of providing a more general account of how realism can be put into practice has been undertaken by Andrew Sayer, from the School of Social Sciences at the University of Sussex. In his 1984 book, *Method in Social Science: A Realist Approach*, he reworks the realist philosophies of Bhaskar, Harré and others, and provides us with an accessible version of a realist project running from general thought to practical implementation.

Systems, regularities and laws

As would be expected from the foregoing examination of realist philosophies, Sayer's work is founded on a critical account of positivism in human geography

Figure 5.5 Two prerequisite conditions for regularity in social science (Sayer, 1984, p. 112)

1. 'There must be no change or qualitative variation (e.g. impurities) in the object possessing the causal powers if mechanisms are to operate consistently. This is termed by Bhaskar the "intrinsic condition for closure". Other things being equal, a clockwork mechanism whose spring suffers metal fatigue will not produce regular movement. Similarly, a pressure group undertaking a political campaign will not produce regular effects if the internal organization of the group disintegrates.
2. 'The relationship between the causal mechanism and those of its external conditions which make some difference to its operation and effects must be constant if the outcome is to be regular (the extrinsic condition for closure). If the political sympathies of the public are changing for reasons independent of the pressure group's campaign, the effect of the latter cannot be expected to be mainfested as a regularity.'

(Sayer, 1985a). One of his key criticisms reflects Bhaskar's description of *closed* and *open* systems in science. Positivism as the search for, and prediction of, empirical regularities seeks to make universal statements and conceptualises both

1. a constant mechanism for causing regularities; and
2. a set of constant conditions in which that causal mechanism operates.

Thus postivism is suited to, and often assumes the existence of, a *closed system*. Realism, however, represents a way of understanding the conditions that must prevail if such regularities are to be produced. Sayer describes two conditions that are a normal prerequisite for regularity within a closed system (see Figure 5.5) and stresses that only if both the intrinsic and extrinsive conditions are met will a closed system exist in which regularities can be produced. In most cases, however, these conditions are violated, and therefore any regularities produced within what are very rarely – if ever – truly closed systems are only approximate and short-lived. One of Sayer's initial premises is hence that social science in general and human geography in particular tend to deal almost exclusively with *open systems*, and so postivistic analysis encounters significant difficulties in handling these open systems in an assumed closed system fashion.

As an illustration of these criticisms of the positivistic approach, Sayer (1985a) discusses the so-called urban-to-rural manufacturing shift in Britain: the apparent tendency for cities and large towns to lose manufacturing jobs during the late 1970s and early 1980s while small towns and rural areas experienced corresponding gains in such jobs. In this case, says Sayer (*ibid.* p. 165),

there is a striking, if approximate and transient 'regularity': the larger the settlement the greater the job loss, with the smallest actually showing an increase, albeit small in absolute terms. The 'messy' and transitory nature of the regularity reminds us that it is hardly a closed system – indeed what else could one expect of something like employment change?!

Most researchers have sought to explain the urban-to-rural manufacturing shift in terms of a positivist search for regularity via the testing of hypotheses:

Hypothesis one   The pattern of manufacturing employment was caused by some areas having 'Assisted Area Status', which was instrumental in luring in-migrant firms to the area. Such a proposition is transparently incorrect, as some areas such as East Anglia experienced a growth in manufacturing employment without being granted 'Assisted Area Status'. It was quite possible that regional development grants did influence the inward movement of some firms in some places, but designation of 'Assisted Areas' was not the cause of the apparent regularity.

Hypothesis two   The pattern of manufacturing shift was caused by the availability of female labour in the receiving area. Again, this may not be accepted as a cause of the regularity. Indeed, questions regarding the 'anatomy' of local labour markets are much more complex than this hypothesis suggests. Factors such as the cheapness, malleability and degree of unionisation of the labour force are at least as significant as gender considerations in some locations, and some relocations of manufacturing concerns have involved a re-structuring of production so as to replace workers with machines. So, whereas the availability of a pool of female labour has undoubtedly been one factor in some cases, its importance does not extend to all circumstances, and it cannot be regarded therefore as responsible for regularity.

Hypothesis three   The pattern of manufacturing employment shifts was caused by 'rurality'. After all, this is the inference within the phrase 'urban-to-rural shift', and the significance of rurality can no doubt be supported by statistical association. This is perhaps the most nonsensical suggestion of the three, because it endows rurality with causal powers when it is difficut to understand how rurality (a chaotic conception if ever there was one) *per se* can cause anything (Cloke, 1987; 1989). In true realist vein, Sayer suggests that we need to know what it is about rurality that could lead it to attract manufacturing industry. Thus, the entire concept of rurality needs 'unpacking', so that many inherent and relevant mechanisms or conditions can be uncovered. In the absence of such an unpacking, any association between rurality and manufacturing shift could be interpreted as accidental.

This illustration neatly displays the fallacious use of positivistic assumptions about the world being made up of discrete events that can be measured as specific phenomena. This perspective relies upon a simple association between cause and effect, real and observed, and implicitly assumes the existence of a closed system. It is taken for granted that causes, effects and observations neatly map onto each other, for within positivism there can be no unobservable causes or effects that are not associated with causes co-terminous in space and contiguous in time. Realism presents an alternative by assuming a stratified and differentiated world made up of events, mechanisms and structures in an open system where there are complex, reproducing and sometimes transforming interactions between structure and agency whose recovery will provide 'answers' to questions posed about processes such as the urban-to-rural shift.

Necessary and contingent

Another important part of Sayer's realist thesis is the stress he places on Bhaskar's distinction between *necessary causal powers* and *contingent conditions*, and between *internal* and *external* relations:

> The relation between yourself and a lump of earth is external in the sense that either object can exist without the other. It is neither necessary nor impossible that they stand in any particular relation; in other words it is contingent. It may nevertheless be possible for one object to effect the other – people may break up lumps of earth or be buried beneath them – but the nature of each object does not necessarily depend on its standing in such a relation. By contrast, the relation between slave and master is internal or necessary, in that what the object is is dependent on its relation to the other; a person cannot be a slave without a master and *vice versa*. Another example is the relation of landlord and tenant; the existence of one necessarily presupposes the other.
>
> (Sayer, 1984, p. 82)

This distinction between 'necessary' and 'contingent' is perhaps best explained by analogy, the best loved of which among realists is that of gunpowder. Gunpowder has the (necessary) casual power to explode, but it does not explode at all times and in all places. Whether or not it does so depends on the right contingencies being co-present, in this case the presence of a spark or other form of detonation. As with gunpowder, so in society certain causal powers are argued to exist necessarily by dint of the characeristics and form of the objects that possess them, but it is *contingent* whether these causal powers are released or activated.

Thus, the effects of causal powers depend upon the presence or absence of certain 'contingently related conditions'. For example, in most cases company owners and managers have necessary causal power to sack members of their workforces, and in most cases the workforce has the necessary causal power to go on strike. But neither sackings nor strikes occur all the time everywhere. Each is contingent upon particular conditions – labour unrest, the restructuring of production, plant closure and so on – for the causal power to be exercised. It is important here to place both illustrations in the context of the distinction between internal and external relations. In the case of the managers and employees, there is an internal relationship because employers could not exist without employees, and vice versa. This particular internal relation has necessary causal powers, plus contingent external powers associated with it. Gunpowder on the other hand is not related to anything – it has its own defining causal powers that are realised when certain external conditions are fulfilled (a state contingent upon external relations). We get to the position here, therefore, where we can almost talk about the 'necessary' contingencies required for the causal (necessary!) powers of gunpowder to be realised in an explosion. Here is a case where 'technical' terms and everyday language can suggest conflicting meanings, and we suggest that this kind of semantic confusion can often cause problems in writings on realism in human geography (as with other philosophical approaches). Overall, then, the distinction between necessary and contingent powers can apply to all objects and relations, but the distinction between internal and external relations relates to interconnectivities between specific objects.

Abstract and concrete

'An abstract concept, or an abstraction, isolates in thought a *one-sided* or partial aspect of an object. What we abstract *from* are the many other aspects which together constitute *concrete* objects such as people, economies, nations, institutions, activities and so on' (Sayer, 1984, p. 80).We have so far seen how necessary causal powers are realised through mechanisms and events. It is therefore the realist task to tease out the causal chains that situate particular events within these deep-seated mechanisms and powers. Realism therefore has needed to develop a research strategy in which concepts are interconnected with and inform empirical materials, which in turn are informed by the theoretical categories. It is important here to recognise a further Bhaskarian distinction used by Sayer when striving to put realism into practice, and here he emphasises two requirements:

1. Theoretical categories, so as to 'get at' necessary relations.
2. Empirical study, so as to 'get at' contingent relations.

The realist terminology for these theoretical categories of necessary relations is *abstract theory*. Abstract here does not imply vagueness, but refers to a particular relationship between causal powers and an object of study: Sayer refers here to a *one-sided* or partial aspect of that relationship. Thus, consider owner-occupiers who are mortgagees with the Lampeter, Cwmann and Ffarmers Building Society. There is a necessary relationship here in that the building society under capitalist property laws remains the owner of properties until the last payments are fulfilled, and could take them over and even sell them if payments were not maintained. This is an important one-sided or partial aspect of the nature of these owner-occupiers. But the fact of owner-occupation does not indicate other necessary relations: for example, with how the individuals spend their leisure time, or the degree to which they use services provided by the local authority. These are 'spurious' relationships in the sense that there is no causal link between owner-occupation and leisure or service use (despite the fact that empirical regularities may well be identified between these very factors).

The realist terminology for the empirical study of contingent relations is *concrete research*. Research of this kind is required in order to discover the actual contingent conditions under which the causal mechanisms we are interested in are triggered. Sayer refers here to a *many-sided* object of study. Thus, our owner-occupiers mentioned above are not only mortgagees but also have particular jobs, particular levels of wealth, particular political, moral and religious beliefs, particular forms of shared decision-making and division-of-labour with their partner (assuming the existence of a partner) and so on. From this kind of concrete research, the exercise and effect of particular causal (necessary) relations may then be abstracted. As Sayer (1984, p. 81) concludes,

> the understanding of concrete events or objects involves a double movement: concrete–abstract, abstract–concrete. At the outset our concepts of concrete objects are likely to be superficial or chaotic. In order to understand their diverse determinations we must first abstract them systematically. When each of the abstracted aspects has been examined it is possible to combine the abstractions so as to form concepts which grasp the concreteness of their objects.

Figure 5.6 summarises the relationships envisaged by Sayer between the concrete and the abstract and the relationships between events, mechanisms and structures. Here is a complex social system in which the activation of particular mechanisms produces effects that may be unique to a particular time and

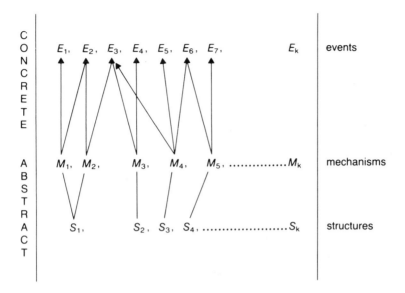

Figure 5.6 Structures, mechanisms and events (Sayer, 1984, p. 107)

space. With different contingent conditions, the same mechanisms may invoke different events, and by the same token the same kind of event may have different causes. The use of abstract theory permits the analysis of objects in terms of their 'constitutive structures', as a part of wider structures, and in relation to their causal powers. The use of concrete research permits an examination of what happens when these combine. Crucially, Sayer (*ibid.* p. 107) adds: 'In the vertical dimension, some readers may want to add a fourth level above events to cover meanings, experiences, beliefs and so forth, but as these can form structures, function as causes or be considered as events I would suggest that they be taken as already included'. So here we have Sayer not only introducing the notion of structure as a shorthand for necessary (and sometimes causal) relations, but we also find him making a half-apologetic disclaimer about certain aspects of human agency that he sees sometimes as forming structures, and sometimes considers as events: and critics of realism may feel that this manoeuvre does not satisfactorily answer the humanist call for people to be taken seriously as people (see Chapter 3). It is important, then, in rounding off an examination of Sayer's portfolio of conceptions of knowledge, that we look a little further at how he deals with the structure–agency debate.

Structure and agency

Whereas Giddens's account of structure and agency as a duality was presented as one of the starting points of his structurationist project (see Chapter 4), realist thinkers have tended instead to allow the philosophy of realism to inform their view of structure and agency. Thus, the debate over the relative roles and natures of structure and agency in social change has been something realism addresses from a particular epistemology and ontology (which, as we have seen, has spawned differing approaches to social science) rather than being something seen as the keystone in the building up of that viewpoint.

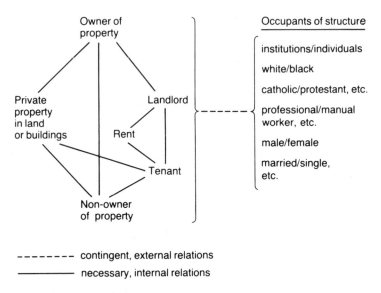

Figure 5.7   Sayer's view of structure (Sayer, 1984, p. 85)

Realist discussion of structure and agency hence sounds more like a series of warnings about the complexity of the issues concerned rather than a deeper engagement with the issues themselves: a situation that demonstrates realism's epistemological involvement with structure and agency, as opposed to the more ontological involvement of structuration.

Sayer, for example, following Harré, defines structures as 'sets of *internally related* objects or practices', and uses the example of the landlord–tenant relationship (which itself is related to factors of private-property ownership, the payment of rent, and the production of an economic surplus) to illustrate structural form (see Figure 5.7). He firmly differentiates between the 'role positions' held by individuals and the actual individuals themselves, stressing that to blame the individual in a particular position for anything that goes wrong is to ignore the process whereby the structure of social relations (along with the existence of associated resources, rules and constraints) may determine the action of that individual in the position concerned, even though these structures will only continue where they are reproduced by human action.

Structures are thereby seen as durable, sometimes capable of causing social change or social conditions, and also capable of locking their occupants into role positions. They are often difficult to displace and transform, yet are continually reproduced by the actions of people, who in turn are often not reproducing structures in any way intentionally. As Bhaskar (1979, p. 44) states: 'People do not marry to reproduce the nuclear family or work to reproduce the capitalist economy. Yet it is nevertheless the unintended consequence (and inexorable result) of, as it is also a necessary condition for, their activity'. This does not mean that actors are seen as cultural dupes, programmed to perform roles and reproduce structures. The actions of actors necessary to reproduce structures in society must be seen as skilled accomplishments born of practical knowledge.

The description of structure and agency by Bhaskar and Sayer draws heavily from similar critiques of Marxism and humanism as employed by Giddens (see Chapter 4). Social processes are not reducible to the seemingly undetermined actions of individuals; but the stress on how action is located within social relations and constrained by conditions that are not of the actor's choosing, tends to ignore the skills and interpretative qualities of actors, and in so doing dehumanises social science. Thus, both agency-orientated and structure-orientated social theories are criticised, but in contrast to the critique in structuration they are not abandoned. Realism caters for the power of these 'two sociologies' as modes of abstraction, but recognises the danger of using them to provide concrete descriptions that solely refer to this theorised knowledge and not to necessary powers and contingent relations in a particular situation. Marxist and humanist theory may both be significant inputs to realist research, but they should not be the end-product of that research.

How, therefore, can we characterise the position of structure and agency within the realist project? Realism's epistemological challenge is to allow for varying relationships between the necessary and the contingent. These relationships can (but do not always) represent the varying interconnectivities between structure and agency, but the warnings are that the causal powers of structures will not always be realised and that not all contingencies will be connected with human agency. (Indeed, the latter-mentioned contingencies may revolve, as it were, around the specific times and spaces in which social actions occur, and not just spring from the 'ambiguities' of human agents. And it is this very treatment of time and space that has proved to be a point of important criticism for Sayer's practice of realism. He tends to view space only at the level of contingency, whereas others feel that space is also implicated at the level of necessity and would therefore insist that, for example, capitalism can be said to have a geography that is fundamental rather than incidental to its very workings. And the same would be said by these analysts for the geographies 'written into' class, gender and other core areas of inquiry; see below.)

Despite these qualifications it is claimed that a realist epistemology is capable of accommodating existing social theories such as humanism and Marxism: these theories arguably identifying two different ontological domains that can, none the less, maintain their integrity as *complementary* parts in the differentiation of social reality. As a result, realism has been associated with Marxist views of historical materialism (see, for example, Williams, 1981), given that the causal (necessary) mechanisms that have been central to realist analyses of society have usually referred to systems of social practices. It should be stressed that not all realists are Marxist in ontology, however, and indeed that by no means would all Marxists accept a realist epistemology (for example, see Harvey, 1987b). Equally, realism has been associated with some of the ideas of human agency stemming from humanist ontology, in that it consistently warns of the need to view individuals as capable and knowledgeable human subjects living in a complex world of meaning, construct and interpretation. Again, however, not all realists are humanists and vice versa.

Sayer emphasises that in real life we live in a world of 'conjunctures' with arbitrary boundaries that haphazardly cross-cut structures and causal relations. This world is pre-interpreted by the people who live in it, and these existing interpretations are fundamental to any attempt to explain the world.

This leads to an interesting duality that marries with Giddens's 'double hermeneutic' or dialogue between insider and outsider interpretations (see Giddens, 1976; Gregory, 1982c; Keat and Urry, 1982; and also Chapter 3):

1. Most individuals, for most of their day-to-day lives, do not make sense of the world by way of the rather clinical abstractions required by realism. Rather, people make their own interpretations, and these may be thought of as *lay or expressive constructions* of reality.
2. Yet, explanatory power is lost if these real-life interpretations are severed from the explanatory causal relations uncovered by realism, which can be termed *scientific or objective constructions* of reality.

The wish to retain both elements of this double hermeneutic has meant that realism has been increasingly harnessed with structuration theory (Gregory, 1982c), at the heart of which we find expressive constructions of reality supposedly 'informing' matters of agency and objective constructions supposedly 'informing' matters of structure.

Another link between realism and structuration occurs through Giddens's contextual theory of the time–space constitution of social systems. Sayer argues that social systems are a continually changing 'jumble of spatial relations', and claims that even though concrete studies may not be interested in spatial form *per se*, space and time must be taken into account if the contingencies of concrete events and the resultant differences to their outcomes are fully to be understood. The importance of space in both concrete research and abstract theory will be discussed later in the chapter. But for the moment we have a sufficient grasp of the realist portfolio of distinguishing concepts – open and closed systems, necessary relations and contingent conditions, internal and external relations, abstract theory and concrete research, and structure and agency – to permit us to understand the importance of Sayer's attempt to apply realism in practice.

Realist practice in human geography

In most discussions of method, the basic aims of social science are taken for granted as the development of a 'scientific' objective, propositional knowledge which provides a coherent description and explanation of the way the social world is. I shall call this the *orthodox conception of the aims of science*. But if we pursue the question of difficulty far enough, there comes a point where we have to reassess these aims and ask whether they generate unreasonable or contradictory expectations. I shall call the alternative the *critical theory conception*. When we throw open the whole question of what social science and related kinds of knowledge are *for*, the difficulties become more comprehensible. What is more, some of our judgements about what are problems and what are solutions have to be reversed.

(Sayer, 1984, pp. 211–12)

Sayer's project of putting a realist epistemology into practice has been radical and innovative, representing an attempt to provide a critical reappraisal of what knowledge is for, and when certain types of knowledge can inform different notions of the problematic. The pattern of realist logic that underpins the practice of realism is familiar given our examination of realism concepts presented earlier in this chapter:

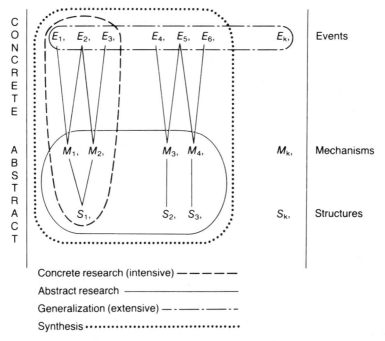

Figure 5.8 Types of research (Sayer, 1984, p. 215)

1. *Events* may be causally explained by reference to *mechanisms*, which in turn may be explained by the structure and constitution of the objects that possess them.
2. Events may be brought about by the conjunction of different causal mechanisms, and may indeed be prompted by the 'accidental' product of the secondary powers that arise from the combination of other objects whose causal powers are independent of the event in question.
3. Explanation of an event is often made difficult by a pre-existing description of that event or series of events (for example, the troubles in Northern Ireland are often viewed as a 'religious' problem, yet other explanatory causes – political, cultural, historical, materialist – can properly be brought to bear on this set of events).
4. Concepts in society may be explained in a way that the individuals concerned themselves understand or interpret them, as well as by reference to their abstract necessary causes and enabling contingencies. However, some concepts which we might legitimately use might be refused or not understood by actors. Thus interpretative understanding of concepts is different from more causal explanation.

Sayer uses this set of explanatory procedures to clarify the relationships between different kinds of research. Figure 5.8 reprises Figure 5.6, but indicates how events, mechanisms and structures can be combined in different ways through research, generalisation and synthesis. Thus *abstract theoretical research* deals with structures and mechanisms, and events are only considered as possible outcomes. Theories of social class that define class in terms of internal relations or social structures are an example of such research. *Concrete practical research* on the other hand deals with actual events and objects, treating them as phenomena that have been brought about by specific mechanisms and structures (each of which will have been indentified and examined through abstract research). Abstract and concrete research should be contrasted with *empirical generalisation*, which merely seeks to establish regularity at the level of events. Finally, it is possible to conceive of *synthesis research* that combines all three of the previous research types in attempting to explain entire sub-systems. Sayer's (1984, pp. 216–17) conclusion from this discussion is that

> the functions of these different types of research are often misunderstood both by users and critics. In particular, researchers often over-extend them by expecting one type to do the job of the others (looking at this from the opposite direction it can alternatively be seen as a form of reductionism),

and he isolates several different ways in which particular research types have been over-extended:

1. *Over-extended abstract research*, where abstract theories are somehow expected to explain events directly (without recourse to empirical research on relevant contingencies). For example, many researchers of the capitalist state have suggested that the concrete nature of state activities can be derived from the most basic categories of (usually Marxist) theories, and have ignored the possibility of developing new abstractions from first studying concrete objects.
2. *Over-dominant abstract research*, where abstract theories are assumed to pre-empt the specific characteristics of the concrete. When (such as in the expectation that class concepts will be exactly reproduced in concrete studies of the population) it is discovered that such pre-emption does not apply in practice, the abstraction is prone to rejection (thus class does not exist) because the shape of contingent relations has been ignored.
3. *Over-extended concrete research*, where the findings relating to a particular contingent part of the system are subject to unwarranted overgeneralisation – for instance, it always rains in Manchester; all southerners are 'yuppies'; there is a 'north–south divide'.
4. *Over-dominant concrete research*, where localised unique findings are given undue prominence without reference to the broader necessary relations that have brought them about. This is a potential problem in some cultural and anthropological studies based on a very small number of 'samples' or 'case studies'.

These various problems will each deny the proper practice of research informed by realist epistemology. The key requirement isolated by Sayer, therefore, is how to conduct *theoretically informed concrete research*, and he proceeds to address this problem in a discussion of *intensive* and *extensive* techniques in research design.

Intensive and extensive research

Concrete research from a realist perspective must account for the complicated nature of the 'individual' (the individual event, phenomena, person or whatever). Therefore, factors of complexity, context-dependency and qualitative change are all relevant features of concrete individuals who are 'identical' in many important respects, and therefore – and as Harré (1979) suggests – by increasing the number of properties used in the definition of an individual,

	INTENSIVE	EXTENSIVE
Research question	How does a process work in a particular case or small number of cases? What produces a certain change? What did the agents actually do?	What are the regularities common patterns, distinguishing features of a population? How widely are certain character- istics or processes distributed or represented?
Relations	Substantial relations of connection	Formal relations of similarity
Type of groups studied	Causal groups	Taxonomic groups
Type of account produced	Causal explanation of the production of certain objects or events, though not necessarily representative ones	Descriptive 'representative' generalizations, lacking in explanatory penetration
Typical methods	Study of individual agents in their causal contexts, interactive interviews, ethnography. Qualitative analysis	Large-scale survey of population or representative sample, formal questionnaires, standardized interviews. Statistical analysis
Limitations	Actual concrete patterns and contingent relations are unlikely to be 'representative', 'average' or generalizable. Necessary relations dis- covered will exist wher- ever their relata are present, e.g. causal powers of objects are generaliz- able to other contexts as they are necessary features of these objects	Although representative of a whole population, they are unlikely to be generalizable to other populations at different times and places. Problem of ecological fallacy in making inferences about individuals. Limited explanatory power
Appropriate tests	Corroboration	Replication

Figure 5.9 Sayer's summary of intensive and extensive research (Sayer, 1984, p. 222)

we decrease the number of individuals who have all of these properties. There are consequently two major options for concrete research, which Harré terms respectively *extensive* and *intensive*:

1. To study a large number of individuals, but to restrict the number of properties used to define them. Extensive research offers some representativeness of wider society, but is liable to exclude essential properties capable of influencing the behaviour of individual people.
2. To study a large number of properties of a small number of individuals. Conversely, this intensive research may ignore many significant parts of the system, as well as many significant individuals.

Sayer's point is that the importance of these two types of research is much more complex than mere questions of scale or depth versus breadth, since the two research designs 'ask different sorts of question, use different techniques and methods, and define their objects and boundaries differently' (Sayer, 1984, p. 221).

Figure 5.9 gives expression to some of those complexities and permits us to build up a profile of intensive and extensive methodologies. Intensive methods beg the question of how some causal process work out in a limited number of cases, and adopt particular research techniques – qualitative methods, participant observation, informal and unstructured interviews, life-histories, ethnographies (see Chapter 3) – that permit the detailed study of the individual in his or her or its causal context, so as to establish interconnections between the necessary and the contingent. The groups to be researched using these methods should be such that the members of the group relate to each other either structurally or causally, thus permitting analysis of causality through an examination of actual connections. The 'informality' of research techniques required in these contexts stands a better chance of 'getting through to' those particular circumstances that are significant to individuals, and permits the corroboration of evidence to ensure that the findings really do apply to those individuals actually studied.

Extensive methods on the other hand do a very different job. They ask the more common question of whether there are general properties and patterns to be discovered over the whole population, and adopt techniques such as large-scale formal questionnaire surveys and descriptive and inferential statistical analyses to uncover these regularities. Evidently, then, in this case pattern and regularity are the guiding indicators of causality. Extensive methods should therefore be employed on taxonomic groups (i.e. those that share formal attributes, but that do not necessarily connect or interact with each other) and will take the quasi-scientific form of asking exactly the same questions to all respondents so as to make 'valid' comparisons devoid of interviewer bias. In this way, explanatory power is sacrificed so that representativeness may be achieved, and so that replicative tests can be undertaken to see how general the findings are in the context of the wider population.

Both extensive and intensive research methods have inherent weaknesses. Extensive research is the weaker explanatory tool so far as concrete events are concerned: it lacks sensitivity to detail, it will not permit the identification of causal mechanisms and, by favouring generalisation over abstraction, it is susceptible both to chaotic conceptions and ecological fallacies. The notion of a *chaotic conception* has already been referred to above in relation to ascribing causality to a category that has little or no internal logic or structural interaction (in this case, rurality), and another example – and one used by Sayer – is that of 'the service sector', which, when unpacked in proper realist style, can be seen to cover an enormous variety of activities that neither form structures nor interact in a causal way. In fact, different components of the service sector (for example, the stock market, a self-employed child-minder, the ice-cream (wo)man and the Christian bookshop) can sometimes appear to lack anything in common, apart from vague references to employment, labour and the transfer of money. *Ecological fallacies* meanwhile refer to the related problem that arises from making inferences about the characteristics of an individual from the characteristics of a group: for example, the inference that high crime rates in city areas with high proportions of black residents means that black people commit more crimes than white people (it could, of course, mean that black people are the victims of more crimes or that crime has no necessary link whatsoever with ethnicity).

The main weaknesses of intensive research methods are that they lack representativeness and may therefore be susceptible to the problems of over-extension of concrete research described above. Care is required here, however, not to under-extend these research methods. Sayer (1984, p. 226) stresses that, although concrete events may be unique, the abstract knowledge of structures into which individuals are locked by specific mechanisms 'may be more generally applicable, although it will take futher research to establish just how general they are. In some cases the unusual, unrepresentative conjuncture may reveal more about general processes and structures than the normal one'. It is also important to underline that as social structures exist on a variety of levels (from the international to the interpersonal), intensive studies need not be local in scale, although they usually have been carried out at this local level.

Sayer's conclusion to this analysis of intensive and extensive research methods is that both are needed in concrete research, although at the time of writing (1984) he considered that intensive methods were undervalued. As the 1980s progressed intensive methods have received greater attention in and beyond human geography (see Chapter 3), although there is at least a suspicion that this is merely a case of using more traditional extensive research with a bit of ethnography 'tacked on' for effect.

Practising critical research

Sayer's integrative treatise on social-science methods from a realist perspective ends with a reflection on how a greater acceptance of the demands of critical theory might be placed alongside those of a realist epistemology. This juxtaposing of realism and critical theory is significant for at least two reasons. First, it reinforces the discussion, which was briefly reviewed earlier in this chapter, about the potential parallels to be found in the directions taken by hermeneutics, critical theory and realism. Second, it serves to pre-empt criticism of the realist project in terms of its incompatibility with an emancipatory trajectory in research. Sayer suggests that alongside the specific types of research (such as action research; see Institute for Workers Control, 1977), which have specifically been linked with critical social science, there are more conventional ways in which the goals of critical social science might be approached. In particular, he notes the necessity for continued abstract and concrete research, and links the recognition given to critical social science with the unavoidability of continuing to understand 'what is' and 'what should be' and of choosing the most appropriate methods of study accordingly:

It should be noted that critical theory does not simply *replace* research on what *is* with criticism of what is, plus assessments of what *might be* from the point of view of emancipation. It would be a poor critical social science which imagined that it could dispense with abstract and concrete knowledge of what is in society. If certain mechanisms are to be overriden or undermined and new ones established we need abstract knowledge of the structures of social relations and material conditions by virtue of which the mechanisms exist. And for some practical purposes, such as economic planning, a detailed concrete knowledge of the system may be needed too.

(Sayer, 1984, p. 233)

He thereby claims at bottom a central role for realist epistemology as a vehicle in which the compatabilities of different ontologies may be driven to the common journey's end of both explaining and even changing the social world.

Evaluating the contribution of realism to human geography

There are a number of ways in which we might begin to assess the impact the work of Bhaskar, Harré, Keat and Urry, and Sayer has had on human geography. One indicator would be to review the wealth of studies during the 1980s that have avowedly adopted a realist approach to a particular object of study, and have thereby arguably offered new insights to long-standing interests in the subject. There are a host of examples that could be served up to illustrate this point, but three examples might be

1. Allen's (1983) realist approach to landlordism, which looks at the different properties that suggest potential causal subdivisions between different landlords (such as employer landlordism, investor landlordism, commercial landlordism and the like);
2. Sarre's (1987) work on the distribution of ethnic housing in Bedford, where using a realist epistemology he tackles the question of whether ethnic segretation represents the choice of the ethnic minorities themselves or is caused by 'broader' structures of the majority society; and
3. Lovering's (1985) realist investigation of the location of defence industries in the Bristol and South Wales areas, which places regional intervention and the structuring of space into the framework of necessary and contingent explanation.

Another way of reflecting the importance of the realist project to human geography is to consider one particular area of study where the adoption of a realist approach does appear to have led to particularly significant insights into existing issues. A powerful example of this genre is Foord and Gregson's (1986) discussion of the notion of patriarchy from the perspective of a realist epistemology. As they argue (*ibid.* p. 198): 'Theorizing partriachy has emerged as a necessary pre-requisite to understanding the nature of women's subordination. But in order to do this, we need to know what constitutes this new object of analysis. Realist methods can help us in this process of conceptualization'. They therefore set out to establish the basic characteristics, and necessary and contingent relations of 'patriarchy', so as to make it a useful rather than a chaotic conception for understanding gender relations. A number of steps are followed in their argument:

• Patriarchy is gathered around the common component of men's domination of women, and thereby defined in terms of 'a form of gender relations in which men oppress women'. Patriarchy is therefore seen as a specific form of the more general object of gender relations.

- In answer to the question 'what are the necessary relations that constitute gender relations?', it is stressed that all social relations will have gender relations embedded within them, but that gender relations are necessary to the understanding of patriarchy while patriarchy is contingent to the analysis of gender relations. Equally, capitalism and patriarchy are contingently related structures that interconnect in particular places at particular times.
- Gender relations should not just be 'added in' to existing materialist interpretations.
- Traditionally, four necessary and internally related gender relations between women and men have been identified: biological reproduction, heterosexuality, marriage and the nuclear family. Of these, however, only biological reproduction and heterosexuality involve universally necessary relations, the other two being geographically and historically contingent.
- Patriachal forms of biological reproduction and heterosexuality involve men's domination of sexuality, and there is therefore a need for realist-informed empirical investigation of how the causal properties of patriachal gender relations work out in particular times and spaces. This would represent a changed emphasis for the purpose of empirical research by feminist geographers.
- The implications of a realist methodology in this case is that the focus on 'individual and unique instances, localities and relations of biological reproduction and heterosexuality' necessitate the use of the intensive qualitative techniques of ethnography.

This very brief summary of Foord and Gregson's arguments conveys a strong sense in which new directions in the study of patriarchal gender relations have been opened up by using a realist perspective. Indeed, the authors suggest (*ibid.* p. 210) that this approach does present an alternative to previous studies:

> We suggest that attempts to combine, as one object of analysis, capitalism and patriarchy, are not particularly fruitful theoretical projects. This contradicts one of the main objectives of socialist feminism in the 1970's and early 1980's which was to *theoretically* unify Marxism and feminism. However, following our argument it is now apparent that this task required the combination to two very different objects of analysis, capitalism and patriarchy. This result proved unsatisfactory and did not enhance our understanding of gender relations.

The evidence from such thoughtful discussions does suggest that for some geographers realism appears capable of accommodating different ontological abstractions and appears to offer an important contribution to human geography, as to the rest of social science. We should note, however, that Foord and Gregson have encountered very considerable criticism on these matters: many have disputed their discriminations between the necessary and contingent relations involved in the subordination of women, and – while some would agree with the realist claim that such discriminations can and should be made – others dispute the 'grand theoretical' ambitions of Sayer's realism as a characteristic manifestation of *male*-centric theorising (it being seen as a peculiarly 'male' trait to attempt to identify 'big order' in the world, whereas women are perhaps more attuned to seeing and tolerating disorder and incommensurability; see Johnson, 1987).

The incorporation of space

Perhaps the most directly relevant contribution of the realist project to human geography, however, relates to the insights it offers on the importance of space in understanding social systems. The 1980s have witnessed a stream of debate regarding the inter-relations between space and social relations (see, for example, Thrift, 1983; Gregory and Urry, 1985; Gottdiener, 1987; Soja, 1980, 1989). Initially attention was focused on a rejection of positivist searches for general laws of spatial patterning, and grand *re*theorising statements issued forth like so much confetti. An example is Soja's (1985, p. 90) 'transformative' statement on the 'spatiality' of social life: 'Spatiality situates social life in an active arena where purposeful human agency jostles problematically with tendential social determinations to shape everyday activity, particularise social change, and etch into place the course of time and the making of history'. Duncan and Savage (1989) characterise this statement as generalised, nebulous, vague and therefore lacking in utility as a guide to the practice of research.

Our direct interest at this point in the debate relating to the spatiality of social relations is that realist concepts and practices have claimed considerable influence both in the way in which the properties of space are analysed, and in the shaping of practical research techniques with which to inform on these properties. Sayer has again been a leading figure in the examination of space from a realist perspective. He outlines two basic principles about the necessary and contingent relations involved:

1. '[A]bstract social theory need only consider space insofar as *necessary* properties of objects are involved, and this does not amount to very much' (Sayer, 1985b, p. 54). Although space is seen as 'making a difference' Sayer suggests that social theorists such as Marx, Durkheim and Weber have been largely justified in devoting so little attention to space in their *abstract* theoretical writings.
2. But *concrete research* must take account of spatial form even if that research does not have a direct interest in spatial differentiation. The difference that space makes is therefore seen to occur at the level of *contingent relations*:

> The recent interest in space and social theory should not allow us to overlook the questions of how social theory has managed to pay space scant attention without too much trouble and how those theorists who have been preoccupied with space have not been able to say very much about it. However, in the sphere of concrete studies, both the difficulty of such research and its poor record in developing explanations owes a great deal to the failure to consider spatial form.
>
> (*ibid.* p. 65)

These realist principles have been advanced and, in some cases, challenged in two very important directions. In the first place, Gregory (1985a) builds on realist ground to promote the view that human geography can take a central position in the reconstitution of human and social sciences. Arguing that *places* are of central importance in the understanding of spatial structures, that a reconstituted *regional* geography is important because studying the web of human relationships should take cognisance of the particular worldly locations of people's daily lives, and that *spatial structure* is a necessary condition for and outcome of social activity, he promotes a strategy for understanding the role of space in society. This strategy works through the elements in Figure

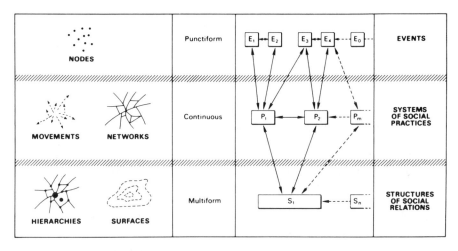

Figure 5.10 Realism and strategies for geographical inquiry (Gregory, 1985a, p. 72)

5.10 and reiterates the realist message that *structures of social relations* are sets of internal relations that have typical ways of acting and can be said to possess causal powers, but here Gregory also wants to insist that these structures – the structures that equate with the necessary properties of 'things' Sayer supposes to be the focus for abstract research – are themselves indelibly bound up with (are dependent upon; are constituted through) their attendant *spatial configurations*.

In fact, Gregory argues that each of the three 'levels' depicted in Figure 5.10 has a 'distinctive spatial structure':

● The spatial occurrence of events is punctiform.
● The mechanisms, or systems of social practice, tend to be routinised and repetitive (and therefore 'continuous') in time and space.
● The structures of social relations are characteristically multiform in their incidence in space.

These three levels of spatial structure correspond in broad fashion to the basic tools of Peter Haggett's (original) *Locational Analysis* (1965) – namely, his 'nodes', 'movements and networks' and 'hierarchies and surfaces' – but, says Gregory, the difference between Haggett's formal geometries and a realist conception of space is that in the latter scheme the levels of spatial structure appear as 'substantive geographies'. In many ways Gregory's project in this respect is no less than to rescue geography from geometry without jettisoning the hard-won insight that spatial structure makes an important difference to the conduct of social life. And it is therefore imperative to grasp the difference between Haggett's geometries and Gregory's substantive geographies, so we quote in full Gregory's own illustration of this difference using a substantive example – one concerned with the spread of machinery through the cotton industry in Lancashire during the early part of the Industrial Revolution – in which he refers specifically to the three levels of spatial structure described above (see Figure 5.11).

Figure 5.11   Gregory's illustration of the substantive geographies of different levels of spatial structure (Gregory, 1985a, pp. 72–3)

A. Each of the three levels shown in Fig. 4 [reproduced here as Figure 5.10] has a distinctive spatial structure. The spatial incidence of events is punctiform, systems of social practice are characteristically routinised and repetitive in space and time, and structures of social relations are typically multiform.

B. These spatial structures correspond, in a very general way, to the nodes, movements and networks, hierarchies and surfaces of Haggett's *Locational Analysis*. But what appeared there as formal geometries appear here as substantive geographies. The difference between the two can be shown by considering – as an example – the spread of machinery through the Lancashire cotton industry in the early Industrial Revolution. If we take each level in turn, then we can construct the following itinerary.

i.   It is possible, at least in principle, to plot the circulation of information about new machinery through advertisements, journals, private correspondence, records of business meetings and ledgers of agents. It is also possible, again in principle, to plot the spread of the machines themselves over space and through time, using insurance registers, business ledgers, Factory Inspectors' reports and the like. We could then draw upon Hägerstrand's diffusion theory to order the innovation process into a sequence in which the incidence of events of one kind (Hägerstrand's 'pairwise tellings') accounts for the incidence of events of another kind ('adoptions').

ii   Both kinds of event dislocated systems of social practice, however, and to put them in their proper context we might draw upon a modified version of Hägerstrand's time-geography to show, *inter alia*, how skilled artisans actively resisted their displacement and subordination through collective projects of their own making (and not out of ignorance, as the classical diffusion model supposes, but because their evaluation of the effects of technical change was sharply different from that of their employers), how the compositions and rhythms of the labour process were transformed and the internal and external linkages of the industry were restructured, how new sexual divisions of labour reverberated through the lives of families and whole communities beyond the factory gates, and how new systems of interaction were spun in space and time as women tramped in to the new mills and their menfolk cast around for other jobs or became overseers and task-masters charged with imposing the new labour discipline on those who worked under them.

iii Diffusion theory is closed around a circle of 'potential adopters'. Time-geography is concerned with the space-time paths and projects of individuals in the population as a whole. *Neither says very much about how the relations between the 'potential adopters' and the population at large are defined and structured.* We know, from the work of Harvey, Massey and others, that possession of and separation from the means of production is of such strategic importance to the class structures of capitalist societies and to the dynamics of capitalist space-economies that we must also examine the distribution of the conditions of production over space, through time and across society. We might, therefore, reconstruct (say) segmented hierarchies within labour markets and the labour process, surfaces of capital concentration and credit availability, and surfaces of accessibility to coal and the carts and canals to carry it.

C. This is not a reductive explanation in which events are reduced to the 'inner logic' of the structures that contain them. Notice that the arrows in Fig. 4 [our Figure 5.10] are double-headed; people are not portrayed as puppets responding to the marionnette movements of the systems and structures which bound them. The hand-loom weavers and factory-masters were not mere ciphers 'standing for' an undifferentiated nexus of capital-labour relations. There were engaged in struggles whose outcomes emerged in the course of daily encounters in the mill or on the street and more protracted battles inside and outside the formal apparatus of the state: Luddism, Chartism and so on. These outcomes had their own geographies and materially affected the systems of social practice and structures of social relations in which they took place. But, of course, the conditions and consequences of these struggles also extended far beyond the vertical array of arenas depicted in the diagram. Some of the arrows are therefore horizontal. Mechanisation within the spinning towns affected (and was affected by) mechanisation within the weaving towns (and on the mosaic of sub-regional differentiation itself); what happened within the cotton industry in Lancashire was not indifferent to what happened within textile industries in other manufacturing districts, and indeed within agriculture in the shires; and it would be impossible to make much sense of the Industrial Revolution in Britain without tracing its involvement with other societies and space-economies within the developing world system.

D. The excavation of all these spatial structures depends upon ordinary language systems. If it 'works', it does so because it draws upon substantive theories of social practice in space and time, uses them to structure complex and fragmented empirical materials, and then in turn 'reworks' those theoretical categories through the empirical grid. Content is poured into the inquiry at every level in a constant and creative process of discovery, and although the same strategy could be used in other contexts (and it could quite properly include some quantitative methods) this is not a formalism: the outcomes would be *different*.

Gregory stresses that his scheme does not involve a reductive explanation whereby events are wholly conditioned by the structures involved. Rather, his arrows in Figure 5.10 reflect the influences that extend in both directions, denoting the causal power of structures, the knowledgeability and actions of human agency, and indeed the struggles that emerged as outcomes (in the example of the Lancashire cotton industry) during day-to-day encounters at the mill and as part of longer-term battles both within and outside of the formal state apparatus. Crucially, however, 'these outcomes had their own geographies and materially affected the systems of social practice and structures of social relations in which they took place' (Gregory, 1985a, p. 73). Figure 5.10 also reflects (via its side-ways arrows) that the struggles and outcomes are interconnected with other conditions and consequences beyond the scope of this (and indeed any other) sub-system of the human environment. Gregory (*ibid.*) draws this conclusion about his attempt to encapsulate different levels of spatial structure:

> If it 'works', it does so because it draws upon substantive theories of social practice in space and time, uses them to structure complex and fragmented empirical materials, and then in turn reworks those theoretical categories through the empirical grid. Content is poured into the inquiry at every level in a constant and creative process of discovery, and although the same strategy could be used in other contexts . . . this is not formalism: the outcomes would be different.

Gregory's approach differs from that outlined by Sayer in that it directly merges realist principles with structuration theory in order to achieve a particular sensitivity to space and time. In so doing he presents a rather more complex account of space than that which emerges from Sayer's strong emphasis on the concrete as the main 'entrance' to the investigation of the difference that space makes.

The second trajectory of advance from basic realist principles towards a more subtle and sensitive appreciation of the importance of space has occurred in the debates over *locality* both as a concept and as a basis for research investigations. There is no scope here for a full discussion of the 'locality debate' (we have mentioned it already in Chapter 2), and in any case this is currently the most obvious case of 'fools rush in where angels fear to tread' in social-science discourse.

Nevertheless, what we can say is that the importance of locality in the context of realism lies in its parentage, its position with regard to abstract and concrete research and the way in which it has been the catalyst for a number of a different 'realisms' of space. First, then, it is possible to trace a lineage directly from realist theory through to locality studies via the work of John Urry. As early as 1981 Urry published a paper on the relationship between regions and social class in which he (following Lipietz, 1980) not only coined the phrase 'locality', but began to flesh it out following his own realist theorisation. Thus, in his view it was the undermining of the coherence of existing regional economies by the turn towards the economising of labour-time through time–space transformations that began to heighten the importance of particular localities, and of the concept of locality in general. In this way Urry's interest in realism worked in tandem with his interest in localities. His 1985 chapter relates locality to *the city*, where he saw a substantial shift occurring in the structuring of different urban localities that previously had been integrated into the production and reproduction of capital. More recently, cities were seen as being reduced to the status of labour pools and so were becoming more integrated with wage-labour in the sphere of civil society rather than in capitalist production. These changing relations emphasise both the importance of particular localities and the way in which the spatial division of labour and the spatial structuring of civil society tend to characterise localites. The early development of the notion of locality as a spatial expression of capital and social relations is thus directly tied in to a prominent member of the school of realist thought and theory.

The second point of importance about localities is that they raise core issues about the relationship between the abstract and the concrete. Neil Smith

(1987b, p. 60), for example, describes the interest in locality as a 'search for the middle ground, an attempt to walk a knife-edge path between polarized excesses of the past; between the abyss of abstract theory on one side and the equally daunting abyss of empiricism on the other'. The question of how and why to undertake studies of localities has thus become implicated in how abstract and concrete issues can be inter-related. Cox and Mair (1989) hence argue strongly that the difficulties inherent in the locality debate result from a conceptual leakage from the abstract–concrete distinction to other dualisms. They suggest that there has been a particular tendency to merge the abstract/ necessary/social/global in opposition to the concrete/contingent/spatial/local (see Chapter 2). Indeed, as we acknowledged at the beginning of this section, Sayer himself stresses the spatiality of the concrete and therefore is open to Cox and Mair's criticism that he thereby condones the restriction of inquiries into the importance of space to concrete analyses of contingent relations. Thus, one of the essential criticisms directed at the theoretical misconceptions claimed to undermine the locality research initiatives (such as the ESRC-sponsored 'Changing Urban and Regional Systems' (CURS) programme) is that these initiatives are encased in the concrete and inaccessible to the abstract. Rectification of this leakage using the proper interconnection between the abstract and the concrete, according to Cox and Mair, will require four important features for future locality research:

1. The assumption that the local must equal the concrete must be broken down. Rational abstraction in combination with empirical study is able to 'derive' particular social and spatial structures that are not necessarily local.
2. Localities in capitalist society are irreducibly political, and therefore will be the arenas of struggle. Matters relating to local politics, especially involving the local state (see Duncan and Goodwin, 1988), must therefore be central to locality research. It might be added here that, while the local state is certainly part of the political process, it is only one part; so the degree to which it is highlighted here in isolation tends to disregard other important political components.
3. There needs to be a strong move away from any idea that a locality is some kind of catch-all notion of a unique place. Local specificity must be addressed via concepts at different levels of abstraction rather than merely explained as a mix of local contingencies. Note again here, however, the difference between this argument and postmodernist claims that the local cannot be explained by anything outside of the local (or, rather, cannot be 'explained' at all in the social-scientific sense of the term; see Chapter 6).
4. Therefore, localities should be conceptualised in terms of hierarchical levels of abstraction, including the transhistoric category of time–space differences (as in structuration) and the socio-historical category of capital relations.

The similarity between these features for proposed 'locality research' and the schema proposed by Gregory (1985a) discussed earlier in this section is remarkable, and again underlines the potential of realist research informed by structuration theory in enhancing our understanding of the spatiality of social life.

A third point of relevance to realism may be rescued from the locality debate, and this is the suggestion that different versions of realism have actually produced different conceptions of locality. Gregson (1987b), in her account of the conceptual duplication involved in locality research, recognises no less than eight common uses of the word 'locality' in social-science research (and this does not include other similar notions such as *locale*; see Agnew, 1987; and also Chapter 4). Gregson argues on this basis that there are several different types of approach within locality research and that none amount to more than a duplication of existing ideas. Without disagreeing with many of her criticisms, it is possible to interpret some of this diversity somewhat differently. There does seem to be confusion within the work of particular authors and research groups as they slide from one concept of 'locality' to another, but underlying this confusion we can perhaps detect a number of different *realisms* of locality. Thus Duncan and Savage (1989, p. 198) speak of Cooke's 'version of realist epistemology' and proceed to suggest that he has 'perhaps misread accounts of realism here' (and presumably this misreading is relative to their own entirely 'correct' reading (?)). Meanwhile, the Duncan and Savage 'version of realism' is itself criticised by Warde (1989, p. 280) on the grounds that 'it can account for spatial variation, but not for any outcomes (courses of action, social processes etc.) that are consequent upon spatial configuration'. We have, in addition, already discussed Urry's and Sayer's realisms of locality. The conceptualisation and practice of locality research is therefore a powerful illustration of how the adoption of realist philosophy and epistemology leads to very different and closely argued practising of social-science research. Whether this is a recognition of over-eclecticism or of a radically new ability to accommodate different ontological allegiances within a single epistemology depends on your point of view, and we turn now to a final summary of the broad criticisms that have been made of realism and that may help us to make up our minds on this point.

Concluding summary of critical reactions to realism

In concluding this chapter we will not attempt to pull together all of the critical strands already introduced, for to do so would be unduly repetitive. Rather, we have tried to convey the basic tenor of the critique that has been levelled at realism: a tenor that appears to engender a defensive response (sometimes aggressively so) when leading proponents of realism are asked to answer their critics (see, for example, Sayer, 1987; 1989a). Part of this defensive posture is an orthodox reaction to unsympathetic critics who give the impression of not having thought too deeply about the selection of grounds for their critique. A case in point would be the claim by Saunders and Williams (1986, p. 295) that 'as the name implies "realism" can be taken to suggest that those who do not follow this method are unlikely to identify the "real" causes of phenomena they are studying', to which Sayer (1987, p. 396) lashes back:

> This is as crass as suggesting that Weberian ideal type methodology appeals to some because it seems 'the ideal thing'. If the authors had troubled to find out what realists actually argued – not only myself but more authoritative writers such as Bhaskar, Harré and Hesse – they would discover that they are insistent fallibilists who reject notions of absolute truth. All knowledge is revisable, they say, including claims about necessity.

In attempting to avoid these less creditable types of criticisms we will restrict ourselves to three major critical themes that have been advanced by human geographers who do appear to have found out what realists have actually argued.

In the first instance, realism is open to claims that it is eclectic, and hence characterised by a lack of identity. As we shall see, those committed to realism might wish to argue the same ground as being *beneficial*, but being tagged as an approach that can be all things to all people (remember the 'we're all realist now' jibe at the beginning of the chapter) carries with it the stigma of person-of-all-trades but expert-of-none. The substance of these matters is that realism as epistemology appears able to accommodate a range of different ontological inputs in a far more compatible manner than the 'pure' abstract theoreticians would wish (see particularly our analysis of structure and agency earlier in this chapter). This is particularly significant in relation to Marxism, with Harvey (1987b) implying that critical realism is incompatible with Marxism, while Saunders and Williams (1986) regard it as a feeble disguise for Marxism and Lovering (1989) sees it as an 'essential foundation for non-reductionist Marxism'. In other words, realism has got various commentators on Marxism in human geography confused about its compatibility with Marxist theory, and perhaps worried too about getting under the realist covers with some theoretically strange companions. Responding to these concerns, Sayer (1989a, p. 223) unambiguously argues for a plurality of theoretical insights into concrete events: 'Concrete research invariably has to mobilise and integrate concepts from several theories, in order to capture the many-sided nature of its object, even where it is a limited problem, such as the explanation of a migration flow', and it is this accommodating embrace to all-comers who bring with them some theoretical explanatory potential that has caused the confusion and worry among part of the intellectual community: and this is either a big beneficial something in favour of realism or a big empty nothing that discredits it in some circles. The whole problem revolves around how we can judge the 'explanatory potential' of a given theoretical standpoint and whether realism provides us with the 'big' criteria for making this judgement. Only by addressing these issues can we tackle the more specific, yet most important, questions of whether realist epistemology is more or less reductionist than Marxist dialectics (for instance) and whether realist ontology breaks away from structuralism or simply translates it into a new language.

We have already pointed out that realism is also seen as offering some compatability with interpretative philosophies of social science, but – and in marked contrast with the Marxist critique of realism – Sayer (*ibid*. p. 207) notes that 'humanistic geographers seem to ignore realism'. Thus, despite Sayer's project of mobilising and integrating concepts into synthesis we have to recognise that realism has been more divisive and challenging to Marxist theorists than with those whose principal focus is on human agency. One reason for this is that humanistic geographers just do not think in the analytic terms of Sayer's realism (see Chapter 3), and so have not been bothered with it (though see Relph, 1989). The breadth of the realist appeal (or the depths to which its eclecticism plummets, depending on how you look at it) also stretches beyond Marxism and humanism, and into those ways of thinking sometimes referred to as postmodern (see Chapter 6). For example, Gregory (1989a) implies that the marriage of realism and structuration forms part of an identifiable path towards certain aspects of postmodernism, leading as they do to the theoretical recognition of diverse 'local knowledges' and 'lifeworlds'; and Lovering (1989, p. 7) sees critical realism as 'sympathetic to those insights which have been assimilated to the less naive versions of postmodernism'. This pathway to postmodernism certainly seems well trodden in most locality research, which began as realist-inspired and has recently been claimed as additionally significant when seen as a feature of the mushrooming 'postmodern spatial paradigm' (Cooke, 1987a; Lash and Urry, 1987).

If we grant credence to the contribution of realism, then, we can see it as an accommodating and synthesising epistemology that asks new questions, prompts a more sensitive analysis and participates in the waymarking of new directions from which to approach human geography. Alternatively, it might be argued that in order to accommodate widely differing ontological and theoretical inputs, realism's elastic epistemology is set at such a generalised level that theoretical contradictions will tend to appear as soon as it is applied to actual concrete situations, for example, in the locality debate. And this latter view conjoins with two rather more specific critical themes.

A second point to make here is that some of those commentators who have compared the task of realism with the methods available for its deployment have concluded that the realist epistemology is *under-equipped* for its task. As John Allen (1983, p. 193) points out,

> Realism is not concerned with each and every social object; it is only concerned with identifying those objects or groups which possess intrinsic causal properties, which, in turn, offer a key to understanding the complex world of social phenomena. It therefore sets itself the analytical task of conceptually specifying such objects, their properties, and their potential range and scope. But it takes this aim upon itself with little in the way of accompanying methodological prescriptions to achieve its goal.

And Allen may have a point here. For all the sophisticated emergence of realism as epistemology, the *methods* made available to social scientists wishing to implement or practise the realist approach basically consist of the sometimes unhappy mix of orthodox extensive research methods (which anyway were previously part of the practice of positivism, the critique of which is very much a foundation of realist philosophy) with relatively untried intensive methods and a 'dash' of abstract theorising. Although qualitative techniques are gaining methodological ground through some accumulated experience of usage, there is still a rather unfortunate tendency to use ethnographies as the subjugated and often exploited partner of more orthodox extensive techniques. As Allen concludes (*ibid*. p. 193), there is still a need to develop further a mode of conceptual analysis and a set of analytical guidelines that 'would enable us to grasp what kind of objects there are in the social world and how they are likely to behave'.

The third and final critical theme to be aired here concerns the relationship between data and theoretical 'truth'. Both Harvey (1987b) and Archer (1987) raised the question of this relationship in the context of a Marxist critique of realism, which goes something like this. Realism proposes a philosophical 'reality' that cannot be reached by conceptual tools such as those made available by Marxist social theory; it further suggests that this gulf between (for example) Marxism and the real can only be bridged in an approximate way, and that the realist project therefore boils down to seeking successively better approximations of the real by continually refining the methodology involved. A number of direct criticisms follow:

- This realist project is viewed as unremarkable, which is the reason why it has become so popular (see the first point above).

- The realist project is easily assimilated into the longer traditions of research in the social sciences, and it could even be viewed as a barely disguised positivism in some respects.
- As realism allows the 'real' to become the ultimate judge of theory, traces of empiricism can be discerned within what is supposedly itself a critique of empiricism.
- There is a danger that theoretical understandings in this context will collapse into a mass of contingencies.
- Moreover, Sayer's conception of the real is bound up with empirical data, and in some senses he appears to be refuting (Marxist) theory on the basis of what he finds in the data: a procedure that again is seen to constitute implicit empiricism.

Such criticisms lead on to important questions regarding how it is determined to everybody's satisfaction precisely what are necessary and what are contingent relations, and how do we know what is a 'rational' or 'chaotic' conception of the real?

Our intuition may by now have suggested to us that Harvey and Archer are not 'big fans' of Sayer's realism. However, they do raise important issues about Sayer's treatise on realist method, and in particular they highlight the criticism that he never succeeds in breaking free from the traditional notions that 'theory is something to be applied to data' and that 'data are the ultimate judge of theoretical truth'. As Archer (1987, p. 392) concludes: 'In my view, then, the essential basis from which Sayer assesses radical research is really nothing more that his own "individual judgement" as to what is or is not important in the data'. Sayer does respond to these criticisms (see Sayer, 1987), but it seems doubtful that the realist epistemology will ever satisfy those human geographers for whom theoretical concerns are paramount. Realism, then, has taught some geographers the lesson of how to unpack some socio-spatial issues, but in so doing has paved the way for what Harvey (1987b, p. 375) has called 'the fragmentations and the cacophony of voices through which the dilemmas of the modern world are understood'. Some will find it unfortunate that realism does not seem to be so helpful in 'repacking' these issues, and so will inevitably resort to their tried and trusted 'packing theories'. Others will entreat us to go further and to celebrate the fragmented realities of postmodern geographies, and it is to these entreatments we now turn.

6

THE DIFFERENCES OF POSTMODERN
HUMAN GEOGRAPHY

Introduction

In recent years a number of human geographers have begun to engage with –
and arguably to contribute to – the debate about what has been called
postmodernism. This is a debate taking place in various intellectual circles: in
particular those concerned to theorise the nature of architecture, literature and
other expressions of a society's culture. As we will see, though, the debate has
spread into arguments about the overall workings of the contemporary human
world and also into reflections upon how we strive to think about these multi-
faceted workings. Some theorists hence identify an *object* of study that comprises these workings – these complex interactions of economic, social, political and cultural processes in the late twentieth-century world – and refer to this
object as 'postmodernity' or as the epoch of 'the postmodern'. Other theorists
are more occupied with developing an *attitude* (or, rather, a range of attitudes)
towards the knowledge we can acquire about the world, towards the methods
we might employ in the process, towards the theories that inform our research,
and towards the way in which we represent our efforts to one another in
words, sounds and pictures. This attitude is that of 'postmodernism' *per se*,
and here we move away from questions about the things actually 'going on' in
the contemporary human world to questions about how we can find out about,
interpret and then report upon these things.

This is an important distinction to make right at the outset, because most of
the geographers who have talked about postmodernism have tended to treat it
simply as an *object*: as a condition of the contemporary world that involves a
distinctive shift in the temporal and (more pertinently) spatial organisation of
economic, social, political and cultural processes. These geographers have
done an exceptional job in beginning to illuminate the 'geography' of
postmodernity, but it is clear that these studies launch from the relatively
familiar perspectives of Marxist geography, humanistic geography or the versions of human geography inspired by structuration theory and realism. These
perspectives have already been discussed in our previous chapters, and – whilst
there are moments of convergence between them and the accounts given by

more obviously postmodern thinkers – there can be little doubt that none of these perspectives are entirely consistent with the attitude of postmodernism as a way of thinking about the human world.

This is because the postmodernist attitude is inherently suspicious of 'grand' intellectual positions such as Marxism, humanism, structuration theory and realism, which claim to be the sole guardians of 'truth'. Indeed, one of the key premises of postmodernism as a way of thinking – one of its few own starting points of any generality – is that we need to contemplate the human world less in terms of 'grand theories' and more in terms of *humble, eclectic and empirically grounded* materials. Rather than thinking in terms of a human world ordered coherently around a clearly defined 'centre', be this a mode of production (as in Marxist thought) or the subjectivity of human beings (as in humanist thought), postmodernism urges us to think more in terms of disorder, incoherence and the lack of a centre determining everything that is 'going on'. Rather than stressing the essential 'sameness' of phenomena that allegedly arises because these phenomena are all expressions of the same centre, the same basic societal organising principles, postmodernism urges a great sensitivity to the 'differences' that exist between phenomena in all sorts of ways both obvious and subtle. And here we meet a second starting point of postmodernism: namely, the focus upon *difference*, an alertness to the many differences that distinguish one phenomenon, event, process or whatever from another, and an insistence on not obliterating these vital differences in the face of grand theoretical statements (whatever their origin). In practice this alertness to difference has many consequences, but in particular it forces us to respect the myriad variations that exist between the many 'sorts' of human being studied by human geographers – the variations between women and men, between social classes, between ethnic groups, between human groups defined on all manner of criteria – and to recognise (and in some way to represent) the very different inputs and experiences these diverse populations have into, and of, 'socio-spatial' processes. At the same time the attention to difference should immediately attract the geographer's interest, given that the discipline has always – except in the most dogmatic corridors of spatial science – displayed a sensitivity to the specific kinds of differences to be found between different (and 'unique') places, districts, regions and countries.

In highly abbreviated fashion the above paragraphs indicate the direction of discussion to be pursued in this chapter, in that we want to describe the related (but not always complementary) manoeuvres identified as the study of postmodernism as *object* and as the development of postmodernism as *attitude*. In the process we will indicate the important contributions human geographers are making, and potentially could make, to both manoeuvres, although we will also emphasise that it is only really the development of postmodernism as attitude that posts a significant challenge to the ways of thinking already explored in this book.

Considering postmodernism as object

The postmodern pop star
Sean O'Hagan, a cultural critic for the *Sunday Correspondent*, had this to say (1990) about the rock musician, Prince:

Figure 6.1   Prince, the 'postmodern pop star'

In effect, he became the first postmodern pop icon, a mischievous, wilfully subversive star who gleefully juxtaposed pop's polar extremes. Sometimes he was Jimi Hendrix, acid rock guitar hero; at other times, James Brown, soul brother, or Little Richard, *outre* rock extrovert. His live shows defied categorisation, merging rock guitar his-trionics with a big band soul revue and some of the most stunning sets and choreo-graphed dancing ever witnessed on a rock stage.

This account of Prince as 'postmodern pop icon' (see Figure 6.1) is actually quite useful for our purposes, since it neatly signals some of the features usually taken as characteristic of postmodernism in today's culture (and indeed in today's society more generally). In other words, it neatly introduces the complex world of postmodernism as *object*. The point is that Prince provides a richly textured 'collage' of music and spectacle (and note too the collage of words in the lyrics of songs such as 'Alphabet Street') in which the audience encounters an assortment of motifs from a variety of different musical and 'showbiz' *genres*: an assortment that is not neatly divided up, like the different sorts of sweets on a 'select your own' stall, but is all mixed together in a manner that is at once playful (in the sense of undermining the coherence and authenticity of particular *genres*) and serious (in the sense that unexpected juxtapositions may shed fresh and revealing light upon the elements being represented to us). The resulting cultural product hence 'defies categorisation', principally because just too many established categories are whirling around in and through Prince's words, sounds and pictures, and in the process the certainties of older pop-music traditions (where certain melodies, rhythms and images connected up easily one to another and perhaps to certain lifestyles beyond) are thrown into question.

At this moment it might be appropriate to ask if Prince is deliberately manipulating his product in the manner just described – is self-consciously adopting a postmodern *attitude* to his records and concerts – and we would suggest that this question would have to be answered in the affirmative, even though it is presumed that he would not employ the academic vocabulary of postmodernism to explain his actions. We thus open a window here on what it means to create something (a record, a concert, but perhaps too a novel, a textbook, a research project) with a postmodern attitude in mind. But this is to run ahead of our argument, for we must first say rather more about the postmodern *object* itself: we need to examine the way in which so many cultural products around us – from small items such as the pop record through to larger items such as buildings and the overall humanly created environments (the villages, towns and cities) containing them – are now interpreted as postmodern. It may seem a lengthy step from one of Prince's albums to the built environment of London's new Docklands development, for instance, but many writers (human geographers included) detect parallels between such small and large phenomena that arise because both are postmodern objects produced by processes stemming from similar roots (which, as we will see, are referred to with terms such as 'the cultural logic of late capitalism').

Postmodern buildings, cities and societies

The term 'postmodern' has been applied to a variety of cultural products, then, and in particular it is possible to identify a range of artistic creations – paintings, novels, plays, films, sculptures and musical compositions – as displaying the fragmentary, 'collage' and 'referential' (referencing numerous artistic *genres* at once) character of a Prince record or concert. Many of these artistic works are deliberately pitting themselves against the character of earlier works commonly referred to as 'modernist', although charting with precision where a modernist movement such as surrealism stops and a postmodern sensibility to art takes over is far from easy (Bradbury, 1988), and at this point it obviously becomes not just a question of the modern or postmodern *object* but one of the modern or postmodern *attitude* to this object (and we will return to this issue in some detail below). We cannot say much about these artistic matters here, even though we recognise that a sensitivity to such matters may actually be of assistance to us as we seek to present, to write, to illustrate, to talk, the findings of our geographical inquiries (see Figure 6.8), but what we do want to discuss is the thread that leads from the art of postmodern architecture into the complex shapes of the contemporary city.

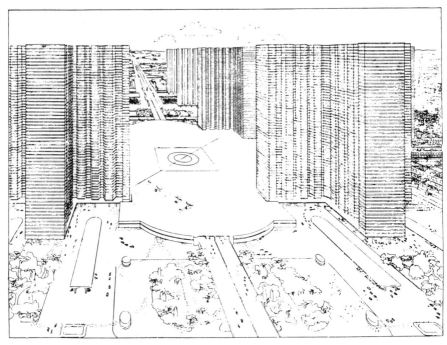

Figure 6.2   Modernist architecture: the 'city of tomorrow', as envisaged by Le Corbusier (1929) (View of the Central Station Flanked by Far Skyscrapers, © DACS 1991)

It is in relation to architecture that the oppositions between modernity and postmodernity become particularly clear, and it is for this reason that many writers – including a geographer such as David Harvey (1987a; see also Dear, 1986) – begin their account of the postmodern with a nod towards the emergence of postmodern architecture from the shell of the modern. As Andreas Huyssen (1984, pp. 12–13) explains,

> Architecture gives us the most palpable example of the issues at stake. The modernist utopia emobodied in the building programmes of the Bauhaus, of Mies, Gropius and Le Corbusier, was part of a heroic attempt after the Great War and the Russian Revolution to rebuild a war-ravaged Europe in the image of the new, and to make building a vital part of the envisioned renewal of society.

This was the architectural vision that in its most extreme guise produced the massive concrete-and-steel constructions in which several generations of Western people (mainly working-class people) were forced to live their lives: this was the vision that produced the great featureless tower blocks Le Corbusier referred to as 'machines for living' and which the architects deliberately modelled on the appearance – and on the supposed 'functionality' and 'efficiency' – of machines, production lines and factories (see Figure 6.2). It was initially an optimistic vision, in which the right design for the built environment was thought to feed directly into the improvement of both human nature and social organisation (it was a definite 'Enlightenment' vision; see below), but, and as we all now know, the reality failed to match the ideal. Indeed, a great many people found it extremely difficult to live happily in these machine-buildings: most people found themselves feeling incredibly 'alienated' from their surroundings, and feeling that human beings were not supposed to exist in these featureless, geometric and artificial (de-naturalised) towers whose simple proportions were so far removed from the scale of the human. The modernist experiment in architecture hence failed because people just could not live with it or in it, and in recent decades the tower blocks have gradually been pulled down:

> Charles Jencks, one of the well-known popularising chroniclers of the agony of the modern movement and spokesman for a postmodern architecture, dates modern architecture's symbolic demise to July 15, 1972, at 3.32 p.m. At that time several slab blocks of St. Louis' Pruitt-Igoe Housing (built by Minoru Yamasaki in the 1950s) were dynamited, and the collapse was dramatically displayed on the evening news.
> (Huyssen, 1984, p. 14; and see also Harvey, 1987a, pp. 260–1)

Modernism in architecture was duly discredited – at least in its more extreme guise, which is not to deny the continued popularity of (say) bungalow designs selected from a 'catalogue' – and in the process it seemed that large and ordered architectural blueprints for how human beings should live their lives were consigned to the dustbin. The question obviously then emerges as to what sort of architecture should today's architects be providing, and it has been in answer to this question that a so-called 'postmodern architecture' has begun to emerge.

There are several strands to this emergence, but (and here it is not difficult to see parallels with Prince's records and concerts) the key manoeuvre has been to abolish the homogeneity and the sheer blandness of modernist architecture, and in so doing to replace the mammoth slabs of gleaming white concrete with an accent on variety, colourfulness, attention to detail and the deliberate mixing of building (and ornamentation) styles from any number of sources. To quote from Huyssen (1984, pp. 14–15) again,

> It has become commonplace in postmodernist circles to favour a reintroduction of multivalent symbolic dimensions into architecture, a mixing of codes, an appropriation of local vernaculars and regional traditions. Thus Jencks suggests that architects look two ways simultaneously, 'towards the traditional slow-changing codes and particular ethnic meanings of a neighbourhood, and towards the fast-changing codes of architectural fashion and professionalism'.

The architectural situation hence becomes extremely fluid and even anarchic, since there are no 'big' rules of style, materials or planning to obey, and as a result there is no reason why the architect of a postmodern housing development in North America should not fuse the latest trends in computer-aided design with classical motifs and the totemic images of indigenous 'Indian' societies (see Figure 6.3). And it should immediately be noted that a postmodern architecture of this kind is of considerable interest to geographers, since an element of what it achieves is to reintroduce a local sensitivity into architecture and to ensure that people are reconnected through their buildings to the deeper traditions and cultures of the places they inhabit: and it is for this reason that David Ley (1989, pp. 243–4) explicitly takes postmodernist architecture as a 'model' that contemporary human geography could follow in its attempt to recover a proper sensitivity to 'people and place'. As a humanistic geographer (see Chapter 3) Ley is also impressed by the concern for 'human proportion' in postmodern architecture, and it is revealing that he sees this

Figure 6.3   Postmodernist architecture: the appearance of a housing project – complete with its references to both warehouses and 'rustic nostalgia' – in the 'postmodern' landscape of Fairview Slopes, Vancouver (from Mills, 1988)

concern as vital in combating all visions – architectural, political, theoretical or whatever – that work on a 'scale' meaningless to human beings living everyday lives in everyday settings (and this is in part why Ley is also critical of 'grand theories' that seem curiously divorced from these everyday human realities; see below). In this respect there is an alliance, albeit a slightly uneasy one, between postmodernism and humanism.

Postmodern architecture is now a crucial component of the built fabric of Western cities, then, and as such it immediately attracts the attention of geographers not just as an inspiration for how they should conduct their own research but also as an object fit for scrutiny by this research. Geographers have long been interested in the architecture of human settlements – consider Paul Vidal de la Blache's (1926, especially Chapter IV) work on the geography of building materials and dwelling types, for instance – and in recent years this interest has intensified, in part perhaps because of the debates over postmodernism (see Knox, 1987; Goss, 1988). An important paper taking seriously the 'architectural geography' of the city is Ley's (1987) account of two rather different 'landscapes' emerging on either side of False Creek in Vancouver, Canada:

> To the north, 'an instrumental landscape of neo-conservatism': high-density, high-rise buildings whose minimalist geometric forms provide the backdrop for the spectacular structures of the sports stadium, conference centre, elevated freeway and

rapid transit system and the towering pavilions of Expo '86. To the south, 'an expressive landscape of liberal reform': low-density groupings of buildings, diverse in design and construction, incorporating local motifs and local associations and allowing for a plurality of tenures, clustered around a lake which opens up vistas across the waterfront to the downtown skyline and the mountain rim beyond.

(Gregory, 1989a, p. 68)

To the north, British Columbia Place displays a modernist architecture, a profit-based development rooted in a 'new right' and authoritarian politics, whilst False Creek Southside displays a postmodernist architecture that dovetails with the liberal individualism and cultural sophistication of its 'new middle class' (professional) residents (see also Mills, 1988, as well as Ley and Olds, 1988).

A not dissimilar investigation of 'landscapes and lifestyles' is also becoming familiar to British geographers, given the amount of effort that has now been expended on trying to tease out the processes currently at work in the restructuring of waterfront sites in British cities, and most notably in London Docklands. Indeed, it is now well known that the landscapes of 'old docklands' – the quays, warehouses, streets and terraced houses – are in many places being swept away to be replaced by the courtyards, apartments, luxury penthouses, shopping and entertainment centres referred to by some writers as elements of the 'postmodern water city'. It is not just a restructuring of the built fabric that is involved here, of course, and John Short (1989) heralds the coming of 'a new urban order' in which the physical obliteration of 'old docklands' by 'new docklands' occurs in tandem with – is both caused by and a cause of – a social recomposition whereby 'yuppies' replace 'yuffies': a recomposition whereby a 'new middle class' of self-confident, well-educated, relatively young individuals (many of whom work in the City's 'financial service' sector) invades the territory of an 'old working class' of depressed and untrained individuals both young and old (although the term 'yuffies' applies chiefly to the young) who no longer have an obvious role to play in London's economic functioning. Short's interpretation of this new urban order pivoting about London Docklands is not without its critics (see Crilley, 1990), and it must be acknowledged that much debate has already been aired over the changing economic, social, political and cultural geography of waterfront communities, but we do not think it contentious to say that investigating such communities signals a move away from the study of postmodern architecture *per se* to the broader study of *postmodernity and the contemporary urban condition*.

And here we evidently reach a claim of considerable generality, and one that goes on to say that the very complexity of the contemporary city 'as a whole' – and not just certain architectural elements within that city – is to be described as 'postmodern'. This is the position a number of human geographers have now reached, and in consequence they acknowledge the fact that the city is no longer organised according to either the simple 'ecological' logic of the Chicago sociologists (it is no longer neatly parcelled up into distinctive 'niches' occupied by distinctive human 'communities') or the simple 'economic' logic of the Marxist geographers (it is no longer fractured into 'suburb-slum' patterns reflecting a straightforward capital-labour 'class' divide). Rather, these geographers acknowledge that the city is – and probably always has been – a much more complex 'socio-spatial' phenomenon, in which all sorts of different people and places are mixed up together in a manner that reflects not the operation of one single logic but the interweaving of a host of different logics (many of which are probably understood but dimly by researchers). Peter Stallybrass and Allon White neatly express this point in relation to the nineteenth-century city, where one dimension 'ordering' the city – the class division, as expressed in the pattern of slum and suburb – was constantly undermined by what another writer calls 'the plasticity of its human geography' (Bailey, 1979, p. 345):

> On the one hand, the slum was separated from the suburb: 'the undrained clay beneath the slums oozed with cesspits and sweated with fever; the gravelly heights of the suburbs were dotted with springs and bloomed with health'. On the other hand, the streets were a 'mingle-mangle', a 'hodge-podge', where the costermonger, the businessman, the prostitute, the clerk, the nanny and the crossing-sweeper jostled for place.
>
> (Stallybrass and White, 1986, pp. 126–8)

The contemporary city is at least as messy and dynamic as its nineteenth-century forebear, and may in practice be more so thanks to 'gentrification' bringing the middle classes back into the inner city (see the essays in Smith and Williams, 1986) and thanks to international migration flows creating multi-cultural urban communities (see the essays in Jackson, 1987), and it is revealing that Harvey – in his *Condition of Postmodernity* (1989b, pp. 3–5) – takes as an index of postmodernity Jonathan Raban's novel *Soft City* (1974):

> The city was . . . like a theatre, a series of stages upon which individuals could work their own distinctive magic while performing a multiplicity of roles. To the ideology of the city as some lost but longed-for community, Raban responded with a picture of the city as labyrinth, honey-combed with such diverse networks of social interacton orientated to such diverse goals that 'the encyclopaedia becomes a maniacal scrap-book filled with colourful entries which have no relation to each other, no determining, rational or economic scheme'.

The city is here depicted as almost entirely without any overall shape or coherence: any sense of a unitary 'community' or of a rationally planned cityscape is seemingly absurd when confronted with the chaotic sites, stages and networks in which, on which and through which the diverse peoples of the city work, rest and play. Moreover, what Harvey also introduces us to through Raban's *Soft City* is the bewildering array of *signs and images* that jump out at us from most contemporary cities; and in this connection we must think not just about the teeming advertisement billboards but also about how all sorts of juxtapositions in the city – consider, as one example, the proximity of a 'cardboard city' for the homeless to a 'glass skyscraper' for the powerful – are unavoidably potent in symbolism. This, then, is the 'postmodern city' that has encouraged such interest in human geographers over recent years.

But what has perhaps been even more noticeable about recent research in human geography is the attempt being made in various circles to document a complex suite of economic, social, political and cultural developments that seem currently to be radiating away from Western cities to energise the changing human geographies of whole societies (see Massey and Allen, 1988; Cooke, 1989a; Hamnett, McDowell and Sarre, 1989; and we might underline here that *rural* areas are not somehow immune to these developments). In this connection some writers are beginning to talk not just about the 'postmodernisation' of buildings and cities but also of the processes involved in the overall shaping – confusing as it may be; and characterised as it may be by variety, by surprising mixes and juxtapositions, by bewildering signs and images – of *postmodernity and the condition of contemporary societies*. To be sure, the vocabulary commonly employed here to represent the complexity of these far-reaching changes does not always speak of 'postmodernism' and, in fact, the terms favoured by many writers are those of 'restructuring', 'post-Fordism' and the 'New Times': terms that tend to accent the economic dimensions rather than the cultural, and that in so doing also indicate the line of reasoning taken by these writers when seeking to explain the changes underway (see below). In October 1988, *Marxism Today* ran a special issue on the so-called 'New Times', for instance, and provided the following (p. 3) assessment:

> At the heart of New Times is the shift from the old mass-production Fordist economy to a new more flexible, post-Fordist order based on computers, information technology and robotics. But New Times are about much more than economic change. Our world is being remade. Mass production, the mass consumer, the big city, big-brother state, the sprawling housing estate, and the nation-state are in decline: flexibility, diversity, differentiation, mobility, communication, decentralisation and inter-nationalisation are in the ascendent. . . . We are in transition to a new era.

The New Times and postmodernity as epoch are fairly interchangeable both as terms and as deeper concepts, and both are designed to highlight the emerging human world of 'flexibility, diversity and differentiation': a world where (to give a simple illustration) a women's worker co-operative making tools for disabled people can co-exist in the same cul-de-sac as a warehouse full of computer games, a social-work office, a row of allotments, a ten-pin bowling club, a mosque and a kiss-o-gram agency.

It is hoped that what we mean by postmodernity as *object* has now become a little clearer, although we should perhaps underline that the crucial point about this particular object – whether we are thinking about pop stars, buildings, cities or whole societies – is that there is no neat single postmodern object awaiting discovery 'out there' in the real world: rather, postmodernity resides more in the complex (and seemingly quite chaotic) collision of *all manner of different objects* in the messy 'collage' of contemporary people and places. The above quote from *Marxism Today* makes precisely this point about the New Times, but it actually does a number of other things as well. Indeed, it begins to reference an argument in which linkages are identified between certain economic phenomena (and notably the appearance of a 'post-Fordist' economic order) and a range of more social, political and cultural phenomena (and in this argument it is only the latter that tend to be accorded the accolade of 'postmodern'). It is important now to offer a brief elaboration of this argument, and in so doing to show how geographers such as Harvey and others end up explaining both postmodernity (or the New Times) and the remarkable 'plasticity' of its new human geography.

Explaining the geography of postmodernity

It as this point that we must turn more directly to David Harvey's thoughts on the *Condition of Postmodernity* (1989b), since it is here that we can gain a clear impression of the argument that human geographers have tended to take when seeking to account for the geography of postmodernity (which is not to deny that there are still important differences between Harvey and other

Figure 6.4   David Harvey's The Condition of Postmodernity *(1989a)*

In this text Harvey provides an account of how postmodernity as *object* is very much the product of certain transformations that have taken place in the workings of capitalism over the last twenty or so years. He argues that the shift from 'Fordism' to 'flexible accumulation' in the organisation of production and consumption has prompted shifts in the 'cultural condition' of Western societies, and that the extreme 'time–space compression' associated with today's capitalist economic order has created a sense of great complexity and even confusion – of 'everything' being interlinked very much in 'the here and now', but at the same time being juxtaposed so that differences between people, places and other phenomena become increasingly apparent – which Harvey interprets as the key experience giving shape to postmodernism as *attitude*. Harvey quite deliberately presents an account in which a cultural 'superstructure' is seen as largely *determined* by an economic 'base', and in so doing restates the importance of Marxist analysis and seeks to deflect the argument that postmodernism as attitude (as a way of thinking) seriously disables the Marxist claim of possessing a privileged insight into the 'truth' of how societies work (see also Harvey, 1987b; Harvey and Scott, 1989; and see our own comments towards the close of Chapter 2). Harvey himself would object to an overly 'structural' reading of his work here. With some justification he claims that his text is concerned less with structures and more with the *processes* – the circulation of capital–binding economy and culture inextricably and reciprocally together in time–space.

geographers concerned with postmodernity, notably Ley, although even here there are as many parallels as there are divergences; see below). Harvey's account is a many-roomed one (see Figure 6.4 for notes on this text), but what he does is to describe postmodernity as 'a historical-geographical condition' – as a condition characterised by a specific 'experience of time and space', and more particularly by the experience of time and space as a messy 'collage' of things all stitched together in the 'here and now' (an extreme case of 'time–space compression') – whose workings can ultimately be explained 'by reference to material and social conditions':

> The experience of time and space has changed, the confidence in the association between scientific and moral judgements has collapsed, aesthetics has triumphed over ethics as a prime focus of social and intellectual concern, images dominate narratives, ephemerality and fragmentation take precedence over eternal truths and unified politics, and explanations have shifted from the realm of material and political-economic groundings towards a consideration of autonomous cultural and political practices.
>
> (*ibid*. p. 328)

Harvey's description here is similar to the above description of the New Times, of course, although Harvey puts more emphasis upon what he sees as the unhelpful 'retreat' in postmodernity from dealing with ethical and political issues, and also upon the 'retreat' in intellectual circles from a concern with the economic 'base' of society towards a concern with its less tangible cultural 'superstructure'. What this means, in effect, is that Harvey regards the overall *attitude* of postmodernism – the overall way of thinking about the world that is now influential in popular, political and intellectual realms (see below) – as itself nothing more than an *object* that can be explained from 'without' using a completely different set of conceptual tools. And what follows from this view of things is the claim that, for all its glitter and apparent novelty, postmodernity is at heart yet another chapter in the story of capitalism, is simply the most recent set of transformations dictated by the logic of the capitalist mode of production, and that as such it is perfectly intelligible to the researcher with a toolbag of Marxist theories: 'shifts of this sort are by no means new, and . . . the most recent version of it [postmodernity] is certainly within the grasp of historical materialist inquiry, even capable of theorisation by way of the metanarrative of capitalist development that Marx proposed' (*ibid.*). Harvey tends to repeat the above-mentioned move of treating the postmodern principally as a matter of culture (as a 'superstructural' matter somewhat ephemeral to the *real* workings of the human world), and this move allows him to adopt a fairly straightforward Marxist analysis in which this cultural condition is seen as fundamentally *determined* by developments currently occurring within the economic 'base' of society. It hence allows him to mobilise a Marxist 'metanarrative' (a big story with a predictable plot and ending; see below) in order to explain postmodernity, and in the process he obviously establishes a line of continuity leading back from his *Condition of Postmodernity* to his earlier experiments as a Marxist geographer (see Chapter 2).

If we examine Harvey's argument in more detail, we can see that what he does is to identify an important shift that has taken place within capitalism over the last twenty years or so, and that has seen the so-called 'Fordist' organisation of capitalist production and consumption replaced by a supposedly more 'flexible' organisation. The accuracy of this identification is contested by some observers (see Chapter 2), and it is also evident that the related debates about 'regimes of accumulation' and 'modes of social and political regulation' are rather involved and often convoluted (see, for instance, the debates of economic geographers recorded in Scott and Storper, 1986; *Society and Space*, 1988), but in essence Harvey suggests that a Fordist era of mass-production and mass-consumption – an era that coincided with class politics and state interventionism – is now being replaced (in some sectors and localities of economic activity) by a rather different assemblage of practices:

> *Flexible accumulation*, as I shall tentatively call it, is marked by a direct confrontation with the rigidities of Fordism. It rests on flexibility with respect to labour processes, labour markets, products and patterns of consumption. It is characterised by the emergence of entirely new sectors of production, new ways of providing financial services, new markets and, above all, greatly intensified rates of commercial, technological and organisational innovation.
>
> (Harvey, 1989b, p. 147)

'On the ground' this shift to a more flexible form of capitalism has often meant the closure of big production-line heavy-engineering plants, for instance, and their replacement, perhaps in the same localities and perhaps even on the same sites, by smaller units that might include the R & D or semi-skilled component assembly divisions of high-tech companies, worker co-operatives involved in 'socially useful production', branch offices of national financial houses, training centres designed to give unemployed people work experience and so on. Slightly more intangibly, this shift has ushered in a dramatic new phase of 'time–space compression' in the world economic order, such that investments, deals and numerous other economic decisions can be transacted between places as distant as London and Tokyo in seconds (thanks to the marvels of satellite communications) and can impinge enormously on the economic geographies of 'an ever wider and variegated space' (thanks to the spread of capitalist practices into more and more regions of the world). This description only hints at the more dramatic dimensions to the shift from Fordism to flexibility, though, and it is vital to recognise that beneath these dramatic 'restructurings' an incredibly rich and detailed set of changes are reshaping technologies, work practices, labour relations, management strategies, and intra- and inter-firm linkages (see the essays in *Society and Space*, 1988).

The conceptual leap Harvey then makes is to regard this deep-seated transformation of economic activity under capitalism as a *cause* of the cultural condition he refers to as postmodernity, and – as should now be obvious – such a manoeuvre means that in the 'final analysis' (in the 'last instance'?) he offers a 'base-superstructure' explanation in which 'the condition of postmodernity' can be read off from the economic logic(s) of 'flexible accumulation'. This explanation proceeds along several lines, perhaps the simplest of which sees Harvey (1989b, p. 344) questioning the extent to which postmodern culture and flexible capitalist strategies should be seen as different anyway: 'I see no difference in principle between the vast range of speculative and . . . unpredictable activities undertaken by entrepreneurs (new products, new marketing stratagems, new technologies, new locations) and the equally speculative development of cultural, political, legal and ideological values and institutions under capitalism'. More specifically, in various places Harvey insists that much of the novel cultural 'collage' demonstrated in postmodern art and architecture is nothing more than an extreme manifestation of the relentless and structurally determined quest for new and unusual *commodities* to sell in the capitalist workplace: as he puts it (*ibid*. p. 336), '[t]he odd thing about postmodern cultural production is how much sheer profit-seeking is determinant in the first instance'. This claim echoes the familiar theme of 'cultural capital', the idea that an individual can 'buy' an exalted place in the capitalist social order by displaying the right sort of cultural attributes (knowledge of the arts; possession of the 'right' educational background) and by acquiring certain marks of personal distinction (the customised car; the individualised home), and it is not difficult to account for many geographical impacts of the 'new middle class' – their 'gentrifying' of inner-city neighbourhoods and colonisation of 'new docklands' – in these sorts of terms.

A third line of reasoning pursued by Harvey proposes a rather more subtle causal relationship whereby flexible accumulation, with its heightened competition between individual producers and with its ever more hungry search for products and markets, calls forth an element of 'cultural mystification' that camouflages these new strategies at a time when the more material panaceas of the 'welfare state' are themselves being eroded. The excitement of postmodern culture, its glamour, its saturation in imagery, its stress on individuality and uniqueness, are hence all interpreted by Harvey as masks disguising both the harshness of these strategies and the callousness of 'New Right' economic policies, and at the same time he detects an 'aestheticising' of poverty that renders the poor, the homeless and the oppressed more objects of cultural interest than objects for radical political action. Closely bound up with this assessment is the role of 'spectacle' in postmodern culture, and here Harvey echoes another familiar theme in suggesting that the glamorous and imposing postmodern redevelopments of many Western inner cities – the 'pavilions' and 'plazas', office towers and apartment complexes, the shopping malls and leisure centres – perform the time-honoured role of 'bread and circuses': 'an ancient and well-tried formula for social control', a formula 'consciously deployed to pacify restless or discontented elements in a population' (*ibid*. p. 88). Whether we are talking about the 'new urban spaces' of Baltimore, New York's South Street Seaport, London's Convent Garden or the 'Metrocentre' in Gateshead, then, Harvey supposes that we are considering a carefully manufactured postmodern spectacle intimately tied into the logic of contemporary capitalism by the very fact that its pleasing and uplifting appearance strives to divert attention away from this logic (although note that the 'imaginary geography' of such spectacles may well be resisted or 'hijacked' for their own purposes by many individual *users* of the facilities, as Shields found in his 1989 study of the famous West Edmonton Mall in Canada). Finally, and as already hinted at, it should be added that much of Harvey's *Condition of Postmodernity* (and perhaps the most difficult and innovative component of the work) is designed to sketch out the connections between the rise of a new flexible capitalism and the dawning of a new experience of time and space upon which postmodern cultural sensibilities are crucially dependent.

Harvey's explanation of postmodernity is indeed a powerful and persuasive one, and there can be little doubt that numerous writers, including most human geographers interested in such matters, pursue their own inquiries in roughly similar directions. In their important text on *The End of Organised Capitalism* (1987), for instance, Scott Lash and John Urry tease out the restless and essentially 'disorganised' character of contemporary capitalism, and in the process they highlight Marx's own view of the fragmentation and transience written into the heart of capitalist societies: the notion of 'all that is solid melts into air' (see also Berman, 1982; Harvey, 1989b, especially Chapter 2). And what these authors also do is to insist that hooked up to today's 'political economy of capitalist disorganisation' is a 'cultural substrate' that can best be conceptualised as one of 'postmodernist culture':

> we are not arguing that there is any one-to-one, reductionist state of affairs in which postmodernist culture is somehow a reflection of the phase of disorganised capitalism. We are moreover not arguing that all or even the major part of contemporary culture is postmodernist. We do, though, believe that postmodernism is an increasingly important feature of contemporary culture, and we intend to try in a preliminary fashion to show how it articulates with some features of disorganised capitalism.
>
> (Lash and Urry, 1987, p. 286)

Lash and Urry then proceed in their preliminary analysis to demonstrate that the imagery and the spectacle of contemporary 'consumer capitalism' are vital in 'predisposing an audience to the reception of postmodern cultural forms' (*ibid*. pp. 288–92), and this analysis leads into some thoughts about how 'postmodern tastes and lifestyle' have clearly become the 'cultural capital' of a 'new middle class' whose own existence is largely a reflection of recent capital 'restructuring' (pp. 292–300). The parallels here with Harvey's arguments about postmodernity as a cultural product of a new flexible capitalism are not hard to discern, and – whilst it is important always to recognise distinctions between the conceptions of different authors – it seems to us that there are further parallels with (say) Fredric Jameson (1984) describing postmodernism as the 'cultural logic of late capitalism', with N. Albertsen (1988) equating postmodernism and 'post-Fordism', or with Edward Soja (1989) reckoning that both 'postmodern and post-Fordist geographies' have been hatched by the most recent 'long wave' in 'successive eras of capitalist development'. Even Ley (1980b, 1987; see also Walker and Greenberg, 1982) – with his humanistic focus upon the creative systems of thought through which different human groups are striving to build new forms of society, politics and culture in the contemporary world – subscribes to a not dissimilar view of things, in that he supposes a key development opening up the possibility for postmodern practices in places like Vancouver is the transformation of society's economic 'base' towards a 'post-industrialism' where the production of commodities has lost its primacy to the production of information (and here Ley draws upon the influential work of Daniel Bell, 1973).

In all of these studies, albeit in their own particular ways we cannot detail here, the postmodern is duly treated as an *object* – admittedly as a very complex object, although usually as a primarily cultural one – and is explained as a product of something else (normally the workings of society's economic 'base', itself normally conceived of as a contemporary mutation of the capitalist mode of production): and what is more, and as will become clearer shortly, in all of these studies the effect is to generate a remarkably *modernist explanation of the postmodern object*! This is not to decry the superb work that has been completed, notably by human geographers with their great sensitivity to the importance of space, place and geography in the condition of postmodernity, but it is to indicate the irony that the majority of these studies analyse the *object* of postmodernity using a way of thinking that owes far more to the *attitude* of modernism than it does to that of postmodernism. And it is to this subtle but highly significant claim our chapter must now turn.

Considering postmodernism as attitude

'Disorder in the windy spaces of my brain'

Some years ago David Sibley published a short paper in the *Professional Geographer* that anticipated the themes running through what we refer to as the *attitude* of postmodernism. In this paper Sibley reflected critically upon the practice of geography as spatial science, not so much on the grounds of its unacknowledged 'ideological bias' as on the grounds that its quest for *order* in the phenomena under study – and a particular sort of order at that: namely, a determinedly spatial order – mirrored a more general 'belief in a unitary cosmic order': a belief that is itself rarely questioned, but that may drastically misrepresent the true nature of the world (and certainly the human world) around us. As Sibley (1981b, pp. 1–2) explained,

> Underlying most writing in locational analysis is a belief in a unitary cosmic order, which we may trace back via Schaeffer and Bergmann to the Vienna Circle. It is a central argument of positivism that order exists in the natural world and that, if we employ the methods of the natural sciences, order can be discovered in social and economic life. Thus Haggett, Cliff and Frey (1977) refer to the spatial order that can be discovered in human geography and, in the first edition of *Locational Analysis in Geography*, Haggett uses the [apparent] order in social and economic life to argue against the idiographic [regional] tradition in geography.

Here Sibley was claiming that geographers such as Peter Haggett merely put a particular inflexion – a spatial inflexion; a concern to discover universal 'laws' of spatial organisation – upon a more widespread belief in the existence of cosmic order, and that in so doing they inevitably turned the discipline towards those philosophies and methods of the natural sciences geared up to (and unquestioning of) the task of learning the physical–chemical laws of natural processes. He went on in this paper to demonstrate how the simple spatial-scientific tool of 'point-pattern analysis' can be used to find spatial regularities in virtually any data set, and he suggested that in the process the messiness of any real distribution being investigated (in this case the distribution of chemists' shops in Leicester) is effectively 'explained away' until we can detect the supposedly 'true' spatial order beneath:

> All those things lying out there and they must fit in. You push and shove the material into the rigid area, getting it in the boundary on one side and it bulges out on another. You run around and press in the protruding bulge, producing yet another in another place. So you push and shove and clip off the corners from the thing so they'll fit, and you press in until finally almost everything sits unstably more or less in there.
>
> (Nozick, 1974, in Sibley, 1981b, pp. 3–4)

Sibley himself has vaguely 'anarchistic' leanings (and here we are talking about an attachment to a definite body of philosophical and political ideas) and as such it is unsurprising that in this paper he displayed such a hostility towards an obsession for order, wherever this may manifest itself.

But his hostility to order here is actually very instructive, and has an importance that stretches beyond a straightforward attack on spatial scientists for finding spatial order where perhaps in reality there is none (or, at least, none that can be meaningfully examined using tools such as 'point-pattern analysis'). Moreover, whilst Sibley initially directed his attack at the natural sciences, it is clear that the claims being made could be extended to various other perspectives that have shaped recent human-geographical inquiries. This possibility was appreciated by Richard Walker (1981), who wrote a response to Sibley's paper in which he sought to defend a Marxist account of the human world against the lines of reasoning which Sibley pursued, and who thereby stressed his own belief in a world ordered according to the economic logic of the underlying mode of production (and more specifically a capitalist world ordered through the scissoring apart of capital and labour in the manufacturing workplace; see Chapter 2):

> Sibley sees only the consciously imposed order of individual industrialists or the state, as I read it. He thus appears to base his notion of order in society on a form of voluntarism. Unconscious or structural forces and outcomes are beyond his ken. He would apparently not be comfortable with Marx's proposition that so long as humankind remains captive to the unconscious forces of history, its behaviour will conform to various historically-specific 'laws'.
>
> (Walker, 1981, p. 7)

Sibley does indeed discuss the imposition of order on populations by industrialists or by the state, and a central contention of his initial paper was to claim that this process can on occasion actually be legitimated by academics with their own obsession for order, and he also demonstrated some awareness of the role that 'deeper' economic and social forces (perhaps bound up with capitalism) might play in this 'making' of order in the human world by human beings. However, he was unhappy about adopting Walker's view that there is a 'bigger' order to the world (a form of 'unitary cosmic order') shaping everything 'going on' in the human realm irrespective of the conscious thoughts and actions of human beings, and acting in such a way that the apparent differences between what is 'going on' in different situations in different times and places are nothing but superficial 'candyfloss' disguising the 'true' nature of things. Sibley himself confessed that Walker may have uncovered genuine weaknesses in his argument, and he rather nicely attributed these weaknesses 'to disorder in the windy spaces of my brain' (1981c, p. 10), but the clear disagreement between Sibley and Walker – a disagreement that indexes *the broader tensions between modernism and postmodernism as attitudes* – arises because Sibley resists Walker's assumptions about the nature of order in the human world and about the privileged ability of Marxism to discover the deepest 'laws' of how this order functions. As Sibley (*ibid*. p. 11) concluded in his reply to Walker:

> My reaction to [Walker] is rather like Stuart Hall's criticism of Althusser [a prominent Marxist theorist], who, Hall asserts, wields a theoretical guillotine to dispatch any concept that has the temerity to stray from its appointed epistemological path. Walker demonstrates greater tolerance than Althusser, but has a tendency to pronounce with an air of certainty on issues that are problematic, even within the restricted field of Marxist social theory.

We would suggest that Sibley would have arrived at a similar response had he been criticised not by a Marxist geographer but by a humanistic geographer intent on stating that there is an order to the world and that it resides solely in (say) the subjectivity of human beings, or by any dogmatic argument claiming to know that there is a 'big' order to the human world that is shaped by such-and-such a set of 'deep' organising principles. And we would also argue that, whilst Sibley's initial paper is not that significant or well-known in itself, it can stand as a useful point of introduction to the matters that we now wish to discuss.

Order, the 'Enlightenment' and modernism

Sibley's paper is potentially very disturbing precisely because it raises a question mark against assumptions so close and dear to most of us, whether we are academics, teachers, students or whatever, that we do not even recognise them as assumptions. This is because most of us have been educated in one way or another to suppose that the world has a fundamental *order* to it, and that this fundamental order can be teased out by the ordered and (to use another significant term in this connection) 'rational' processes of intellectual inquiry. Natural scientists obviously believe that the natural world 'obeys' certain 'laws' (those of Newtonian mechanics, those of Darwinian selection), and so physical geographers – whilst they might debate the relative merits of positivism or critical rationalism as an appropriate philosophy to frame their inquiries (Haines-Young and Petch, 1986), and whilst they may talk about 'chance' and 'chaos' as ways of conceptualising the complexity of physical systems – are in little doubt that there is a 'true' order waiting to be discovered in the soil profile, drainage basin and glacial trough. The assumptions and attendant approaches of natural science have obviously had a great impact upon the social sciences, human geography included, and in many ways the history of social science is the history of borrowings from, and of occasional and inspired moments of resistance to, the ideas and methods of natural science (the shifting of positivist 'science' into the heart of geography as spatial science being just one example). As a result the social sciences have by and large taken on board a natural-scientific notion of order existing 'out there' in the 'real' world, and have hence embarked upon quest after quest for the 'true' order of the human world – for the 'true' way in which human agents interact with one another, with their institutions and with other processes, forces, mechanisms and so on (the terminologies are endless) – and have usually accepted the natural-scientific view that this order *pre-exists* any attempt by human agents to conceptualise it (see the relevant discussion in Chapter 5). It is actually quite surprising to learn just how many intellectual positions within the social sciences rest at bottom upon a natural-scientific model of how the world works and of how we should research it, and this is the case whether we are thinking about approaches informed by the 'Marxist science' of 'historical materialism' (see Chapter 2), by humanistic philosophies such as 'phenomenology' (see Chapter 3) or by various other 'grand theories' of the human world.

But it is not only natural science that teaches us to believe in a fundamental order to the human world, since throughout much of the Western world a crucial horizon of understanding remains that of a God-centered Judeo-Christianity that posits an all-encompassing order to all creation: an order crafted by God, by a single perfect 'centre' to everything, and an order that surpasses human understanding even if suggestions as to its dimensions and meanings can be gleaned by the faithful. Even if this powerful vision is rejected, the alternative scenario of a Godless creation – a creation without order, without centre, without 'rhyme or reason'; a creation that is ultimately nothing – is often resisted not by reinstating God *per se*, but by proposing something else (the 'transcendental essence' of human being; the determinations of the mode of production; the recursive interactions of agency and structure; or whatever) that might be thought by some to 'fill the gap'.

The concept of order as briefly sketched out here, complete with its attendant concepts of truth, reason and logic, is one that can be traced historically in the guise of what is often referred to as the *Enlightenment*:

> The European Enlightenment of the eighteenth century provided one of the essential frameworks for the development of the humanities and social sciences. It was, above all, a celebration of the power of reason and the progress of rationality, of the ways in which their twin engines propelled modernity into the cobwebbed corners of the traditional world.
>
> (Gregory, 1989a, p. 68)

The Enlightenment was thus a movement 'which stressed tolerance, reasonableness, common sense and the encouragement of science and technology' (Edwards, 1988, p. 272), and in some ways it constituted an extension of the Renaissance equation of science and humanism, which made it the duty of human beings to 'exercise their reason' in seeking 'to know God through knowing and perfecting his creation' (Cosgrove, 1989b, p. 194; see also Chapter 3, and note that there are important overlaps between the 'stories' told in both Chapter 3 and the present one). The Enlightenment perhaps relegated God more to the sidelines of intellectual inquiry – although He certainly did not disappear from the scene altogether – but it is not too difficult to see that the Renaissance search for 'the consistency and pattern written by God in creation' (Cosgrove, 1989b, p. 194) continued into the Enlightenment as the search for the various 'laws' underlying the workings of both natural and human worlds. Indeed, it was through the Enlightenment that a law-seeking form of science distinctly *modern* in its appearance became commonplace, and it was the Enlightenment thinkers (the *philosophes*) who 'tried to apply the scientific style of thinking to the regions of aesthetic, social and political theory' (Gay, 1969, p. 126) and who thereby announced that the world of human thoughts, actions and institutions was open for investigation in precisely the same manner as was its natural counterpart.

Moreover, what must always be appreciated about the Enlightenment is the 'emancipatory' impulse that infused it: the genuine expectation shared by many educated people in many different places (and in this respect Gay talks of a 'geography of hope') that the attainment of improved knowledge would lead very directly to human freedom, to a liberation from both the drudgery and the dangers associated with eking an existence out of the natural environment. As the philosopher John Locke put it,

> [There is a] large feild for knowledg proper for the use and advance of men . . . [to] finde out new inventions of dispatch to shorten or ease our labours, or applying sagaciously togeather severall agents and patients to procure new and beneficiall productions whereby our stock of riches (i.e. things usefull for the conveniencys of our life) may be increased or better preserved . . . [F]or such discoverys as these the mind of man is well fitted.
>
> (Locke, in Gay, 1969, pp. 6–7)

These ambitions of the Enlightenment are obviously ones that have been reiterated down the years leading to the present day, and some might see in them a worryingly utilitarian justification for science, but what must also be recognised is that many present-day critical thinkers (including a few humanists, but principally including Marxist theorists working in the shadow of Jurgen Habermas) still hold on to the emancipatory ideals (and hence to other concepts such as order, truth, reason and logic) bound up with the 'spirit' of Enlightenment.

The relevance of the discussion here to a chapter on postmodern geography may not immediately be obvious, and so let us refocus matters by emphasising that 'postmodernism is, in its fundamentals, a critique of what is usually called "the Enlightenment project"' (Gregory, 1989a, p. 68). In other words, it is the Enlightenment obsession with order, truth, reason and logic that a postmodernist attitude is suspicious about (as we will explain in detail shortly) and it is also the case that the emancipatory ambitions of the Enlightenment are seen as problematic; not because the hypothetical ideal of a free human society is not a laudable one, but because the concepts and practices designed to speed us to this ideal are themselves seen as irredeemably flawed.

There is a further equation involved here, of course, in that the project of the Enlightenment is viewed by many as synonymous with that of *modernism* (which is why it is resisted by *post*modernism), although sorting out the precise connections to be made in this respect is far from easy. Some of the difficulty arises because the term 'modernism' is most commonly used to describe a specific movement (or collection of movements) within the arts, and as such has a much narrower meaning than that we conjure up with the term 'Enlightenment'. Malcolm Bradbury (1988, p. 539) writes as follows:

> 'modernism' (or 'the modern movement') has by now acquired stability as the comprehensive term for the international tendency, arising in the poetry, fiction, drama, music, painting, architecture and other arts of the West in the last years of the nineteenth century and subsequently affecting the character of most twentieth-century art. The tendency is usually held to have reached its peak just before or soon after World War I, and there has for some time been uncertainty about whether it has ended.

This tendency is said to have included numerous well-known experimental artistic traditions – symbolism, impressionism, cubism, dadaism and surrealism, to name but a few examples – and it must immediately be acknowledged that these traditions were hardly straightforward exponents of an art preoccupied with order, truth, reason and logic (the role that might be ascribed to the earlier artistic traditions of realism and naturalism). Nonetheless, what these tendencies did share was something of the Enlightenment 'spirit': a belief that through art – as a complement to science (and note how fascinated many of the modernists were with new technology, notably with respect to architecture; see above) – it would be possible to install 'the new', to challenge orthodoxies, to point the way forward to an improved understanding of human nature and society, and to herald radical programmes of human and social transformation.

It is precisely this belief in an 'enlightened modernism' – a form of art which is at the same time a programme for life – that a writer such as Habermas seeks to defend (Habermas, 1981; Huyssen, 1984), and it is from this position that he directs fierce criticism at any intellectual development, whether it occurs in art, science, social theory or whatever, where Enlightenment concepts (order, reason) and ambitions (emancipation, progress) are seemingly dismantled for no purpose other than that of a perverse *negation*. What all of this means is that (in large measure as a result of Habermas's arguments) the idea of modernism has been expanded beyond the books, paintings and sculptures of art to inherit the complex package of concepts and ambitions associated with the Enlightenment (which itself was an inheritor of the Renaissance), and it is this complex package that we will identify as the *attitude* of modernism. And it is this complex package that we suggest is opposed by the *attitude* of postmodernism, which must itself be understood as a complex amalgam of artistic, theoretical and practical (social and political) reactions against all the jumbled elements of modernism.

This claim needs to be approached very carefully, though, since it is important to reiterate that we suppose postmodernism as *attitude* to be more than just an epochal thing, to be more than just the flavour of the moment, since an attitude sceptical about the twin concepts and ambitions of 'enlightened modernism' has been around ever since artists and other intellectuals began to think in terms of order, reason, emancipation and progress. Thus, and as Marshall Berman (1982) shows, even in nineteenth-century literature it is possible to find a sad acknowledgement that both the intellectual search for order and the practical search for emancipation are plagued with problems of disorder, uncertainty and people who 'lose out': and as Berman puts it, drawing here upon the words of Marx, there is a clear recognition that the seemingly 'solid' ideas and promises of modernism will always 'melt into air'. Nevertheless, it is perhaps only in the later years of the twentieth century that the claims of modernism have become subjected to a sustained attack from without – from a range of perspectives that self-consciously stand outside of the art galleries and theoretical writings of modernism – and it is perhaps only now that it becomes appropriate to speak of a widespread and seemingly quite convincing postmodern attitude. And it is perhaps only now that it becomes appropriate to speak of a postmodern attitude permeating a discipline such as human geography, and to acknowledge that such an attitude is capable of destabilising virtually all of the ultimately modernist perspectives that have been adopted by human geographers in recent years: namely, positivist spatial science, Marxist geography, humanistic geography, structurationist geography and realist geography (for a slightly different account of this 'postmodern challenge' to human geography, see Dear, 1988, and see Figure 6.5 for notes on Dear's paper). It might be added, in part as an aside and in part to flag a later suggestion, that earlier approaches adopted in human geography – and most notably the various versions of *regional geography* (see Chapter 1), may not be so disabled by the postmodern attitude, and it is arguably the case that a form of postmodern 'sensibility' has long been present in the care geographers have customarily shown for the specific lands and peoples of specific places.

Before outlining in a little more detail the character of the postmodern assault on modernism, we should first emphasise that – whilst we would accept

Figure 6.5 Michael Dear's 'The postmodern challenge: reconstructing human geography' (1988)

In this paper Dear offers 'a report on geographical knowledge' (a title echoing the subtitle of Lyotard, 1984) and identifies what he regards as a worrying trend towards fragmentation (a 'depressing disarray') where-by human geography is splintering into a host of specialist sub-disciplines – urban geography, agricultural geography, environmental perception and so on – and is entertaining an 'anything goes' stance towards theory. Dear views this trend as symptomatic of a 'postmodern challenge', which questions all stabilities, accepted wisdoms and 'grand theories' within academic inquiry, and then suggests that we need to move beyond the *deconstruction* (see Figure 6.7) urged by postmodern thinkers to a *reconstruction* of human geography designed to embed the discipline more securely in the 'social theory movement' and to establish a new 'internal order' for it. The latter exercise – which insists that 'the central object in human geography is to understand the simultaneity of time and space in structuring social process' (p. 270), and that lays emphasis on the geographical study of economic, social and political processes – supposedly 'faces up to the fact of relativism' in an irreversibly postmodern academic environment, but such an ex-ercise is certainly not consistent with 'a pure postmodernist stance' (as Dear himself acknowledges) and actually takes us back towards the very 'metanarratives' Lyotard (an important guide for Dear earlier in his paper) explicitly rejects.

certain aspects to the postmodern critique of modernism (and more particularly of modernist human geography) – we also share some of Habermas's
concern about the potentially reactionary and 'conservative' nature of this
critique (see also below). In consequence, we want to follow the likes of
Huyssen (1984) and Craig Owens (1985) in finding a 'radical' turn to postmodernism: a radical turn that arises because its sensitivity to *difference* demands an alertness to all of those 'other' human groupings (women, children,
workers, ethnic minorities, gays, disabled people, nomads) who depart from
the modern norms – those of a white, middle-class, middle-aged, male
'straight' society – that frame much of today's intellectual and practical activity

Figure 6.6   Towards a social geography of 'others'

Huyssen (1984, p. 50; see also Harvey, 1989a, p. 113) suggests that
postmodernism as *attitude* possesses a radical edge in its great sensitivity to 'otherness': to 'the multiple forms of otherness as they emerge
from differences in subjectivity, gender and sexuality, race and class,
temporal (configurations of sensibility) and spatial geographic locations
and dislocations'. So much of our theorising and research as social
scientists has remained insensitive to this range of 'others', and this has
been very much the case in the discipline of human geography. As
Gregory (1985a, p. 63) puts it:

> [t])he incorporation of human agency is necessarily *differentiated*. Once we
> accept that people have a place in human geography, then we also have to
> accept that they do not fall out of packets like so many green jelly-babies.
> They are sorted ('structured') by culture and class, age and sex; and yet how
> much of human geography continues to represent the world of the white,
> middle-class, middle-aged man as its 'ideal type' . . . ?'

There are encouraging signs that this blindness to the diversity of human groups is now being rectified, however, and it is possible to identify
an emerging corpus of geographical studies concerned to investigate
the complex associations binding all manner of peoples to the spaces
and places in which they live their lives: and here the concern is with
both *objective* circumstances (the external conditions shaping the lives
particular groups lead in particular homeplaces, workplaces and public
places) and *subjective* dimensions (the internal understandings and
emotional attachments particular groups hold with respect to their geographical surroundings; see Chapter 3). It is hence possible to talk of an
emerging social geography of the 'other', a development neatly paralleling developments in social history, cultural anthropology, archaeology
and cultural studies. The rise of *feminist geography* has been crucial in
forcing the discipline to take more seriously the heterogeneity of its
human subject-matter (see Women and Geography Study Group of the
IBG, 1984; Little, Peake and Richardson, 1988), but so too has the
appearance of geographical work dealing more 'theoretically' and
'socially' with issues of race and racism (see Jackson, 1987) and with
all sorts of 'other' minority groups or, as Sibley calls them, 'outsiders'
(see Sibley, 1981a; Philo, 1986; Winchester and White, 1988).

in the West (see Figure 6.6. for an elaboration of this claim in relation to social geography). Furthermore, we are attracted to the suggestion that combining this distinctively postmodern sensitivity to different *peoples* with our long-standing geographical sensitivity to different *places* signals one (though not the only) road to travel in forging a valuable and still *critical* human geography for the 1990s. Let us now begin to expand upon these various claims.

From modernist 'metanarratives' to postmodern human geographies

Various contemporary thinkers could be mentioned in this connection, and we could begin by explaining what is supposedly involved in the postmodern 'technique' of *deconstruction* favoured by North American literary critics (see Figure 6.7 for notes on this technique), but it is perhaps from the ranks of French intellectuals that the most decisive and influential arguments have originated. The most explicit statement is that of Jean-Francois Lyotard, whose text on *The Postmodern Condition* (1984) provides an account of how 'the condition of knowledge in the most highly developed societies' has – as a result of the computer-aided explosion in both the amount of information available to us and the variety of forms in which it arrives (the variety of 'language games' or 'discursive species') – shifted irreversibly from a relatively simple modern state to an incredibly complex postmodern state. He suggests that the modernist approach to knowledge tends to think in terms of 'science', whether it is dealing with natural or human phenomena, and in terms of 'scientific

Figure 6.7　'Deconstruction'

Deconstruction is first and foremost a 'technique' for analysing texts – primarily philosophical and literary works – and as such it was introduced by Derrida in his 1967 text *Of Grammatology*. As Poole (1988, p. 206) explains:

> Derrida showed that, by taking the unspoken or unformulated propositions of a text literally, by showing the gaps and supplements, the subtle internal self-contradictions, the text can be shown to be saying something quite other than what it appears to be saying. In fact, in a certain sense, the text can be shown not to be 'saying anything' at all, but many different things, some of which subtly subvert the conscious intentions of the writer.

This technique for teasing out the incoherencies, limits and unintentioned effects of a text – an exercise that thereby scuppers any simple notion of being able to discern the real 'meaning' of a text (a notion that privileges the author over the reader) – has had an enormous influence, notably upon North American literary theorists, but what has also happened is that the principle of deconstruction has been extended to the analysis of the incoherencies lying at the intersections of all manner of texts, cultural products, social settings or whatever. As a result numerous researchers, human geographers included (see Dear, 1986, 1988), now talk about 'deconstructing this, that or the other' when they are not so much pursuing a line of inquiry inspired by Derrida as simply highlighting aspects of a given theory or real-world situation that do not seem to 'add up' or 'fit together' all that well.

enterprises' that anchor themselves in particular sets of 'big' and 'deep' claims about how the world operates (and it is in this connection that claims about order, truth, reason and logic emerge): 'I will use the term *modern* to designate any science that legitimizes itself with reference to a metadiscourse . . . making an explicit appeal to some grand narrative, such as the dialectics of Spirit, the hermeneutics of meaning, the emancipation of the rational or working subject, or the creation of wealth' (*ibid.* p. xxiii). We hence encounter Lyotard's notion of the 'grand narrative' or 'metanarrative', and – whilst his use of terms such as 'the dialectics of Spirit' may be rather unfamiliar to us – what he is actually doing here is to identify versions of Marxism and humanism (alongside neo-classical economics) as metanarratives that are thoroughly modern and therefore open to criticism on several counts. And he begins to signal the basis for these criticisms in this, admittedly another somewhat difficult, passage:

> I define *postmodern* as incredulity toward metanarratives. . . . The narrative function is losing its functors, its great hero, its great dangers, its great voyages, its great goal. It is being dispersed in clouds of narrative language elements – narrative, but also denotative, prescriptive, descriptive and so on. Conveyed within each cloud are pragmatic valencies specific to its kind. Each of us lives at the intersection of many of these. However, we do not necessarily establish stable language combinations, and the properties of the ones we do establish are not necessarily communicable.
>
> (*ibid.* p. xxiv)

What Lyotard is getting at is, as already implied, the fact that we are all now confronted with *messages* 'coming at us from all sides' in a bewildering variety of forms (a news bulletin on the television; a love song on our record-player; a package on our micro; a letter; a memo; a novel; a textbook; a video; a conversation), and that in the process any over-riding sense of order, truth and the suchlike is effectively 'dispersed' among a host of different 'clouds' of communication. Each of these clouds possesses its own particular way of communicating and can tell its own particular 'truth', but it appears that few of the clouds are strictly complementary and that we therefore tend to be bombarded with competing messages (and so tend not to exist in the middle of 'stable language combinations'). Lyotard's thoughts in this connection are actually very similar to those of Habermas, but whereas Habermas supposes that there is still some possibility of establishing meaningful communication between the many clouds – some possibility of saving 'enlightened modernism' from dispersal among the clouds – Lyotard can see only a fragmentation that leaves the old metanarratives (and also any attempt to construct new ones) in a hopeless position.

It is vital to grasp the further point that Lyotard is not only suggesting that the sheer diversity of messages and languages currently bombarding us militates against any possibility of thinking in terms of metanarratives, but he is also suggesting that we encounter in this diversity the *real* 'truth' of the human world – the only 'ontological' claim of any generality to be made about this world – that it is constituted of nothing but fragmented clouds of communication bound together by nothing but garbled dialogues between peoples who do not (who cannot) properly understand one another. Any theoretical account seeking to impose an order on this chaotic reality, whether it be a Marxist account talking about the universality of class interests or a humanist account talking about the universality of human nature, is hence viewed by Lyotard as a *modernist* endeavour sadly out of step with the reality of the *postmodern condition* beyond (which demands an entirely different intellectual response; see below): 'the society of the future falls less within the province of a Newtonian anthropology (such as structuralism or systems theory) than a pragmatics of language particles. There are many different language games – a heterogeneity of elements. They only give rise to institutions in patches – local determinism' (*ibid.*). A 'Newtonian anthropology' – a science of society that specifies fundamental and universally valid ordering principles in the same way as do the laws of Newtonian physics – is rendered impossible by the great diversity of different ideas, thoughts, hopes, fears and ways of communicating possessed by very different groups of people. To put matters like this is perhaps to extend Lyotard's argument in a more anthropological and sociological direction than he would have intended, in that we have related Lyotard's 'different langauge games' to different human groups (to differences that exist between groups on the grounds of gender, class, race, age, health, lifestyle and so on) and in so doing we are insisting that modernist metanarratives are seriously compromised by the profound *differences* that fracture human populations into so many groups and sub-groups. But what we can also say, and this is to put a more geographical inflexion on Lyotard's own words, is that a 'Newtonian anthropology' is made doubly difficult by the diversity introduced into the analysis by the simple fact of different peoples living in very different places: by the simple fact that different peoples make very different lives for themselves depending on the particular places in which they work, rest and play (and hence Lyotard's observation about those 'institutions in patches' produced by the forces of 'local determinism'). Bringing these arguments together, we hence can conclude that modernist metanarratives fall apart when confronted with – and are worryingly insensitive to – the differences between different peoples and different places, whereas a properly conceived postmodernist approach to academic inquiry remains ever alert to the subtleties of what Clifford Geertz (1983) calls the 'local knowledge' of particular peoples in particular places (see also Rose, 1988 and the relevant discussion in Chapter 3).

Our claims about the contours of a properly conceived postmodern human geography should already be becoming clearer, then, but we need to underline even more forcefully the extent to which a postmodernist attitude challenges the ordering propensities of modernist metanarratives. And it is perhaps in the work of Michel Foucault, another French thinker, that we can pick up a line of reasoning that speaks most directly to our concerns as human geographers. We say this because at one point Foucault mounts an attack on all of the metanarratives that are conventionally employed by *historians* to bring order to the seeming chaos of historical phenomena, events and processes, and in so doing he signals the need to engage directly with this chaos instead of striving to smooth it out under the influence of some neat but ultimately untenable 'grand theory'. In his text entitled *The Archaeology of Knowledge* (1972, p. 9), for instance, he announces his dislike for what he terms 'total history' (not to be confused with the more specific *histoire totale* practised by the 'Annales School' of French historians):

The project of total history is one that seeks to reconstitute the overall form of a civilisation, the principle – material or spiritual – of a society, the significance common to all the phenomena of a period, the law that accounts for their cohesion – what is called metaphorically the 'face' of a period.

Foucault is here referencing all manner of historical inquiries – positivist, Marxist, humanist, structuralist or whatever – all of which he is deeply critical of because of their desire to impose an a priori and often rather grand interpretation (a 'totalising' interpretation) on the chaos of history itself. For Foucault, all such impositions are suspect because they introduce a measure of order that remains stubbornly alien to the detailed confusions, contradictions and conflicts that have been the very 'stuff' of the lives led by people in the past. They operate through imagining a 'central core' to the human world – a core that might involve the words and deeds of great heroes, the cultural traditions of a given society or the workings of capitalism (to name but a few possibilities) – from which a chain of influences supposedly radiates out to govern everything that is 'going on' in the civilisation or period under study. For Foucault, all such total histories are anathema because they stand squarely in opposition to his own belief that 'nothing is fundamental: this is what is interesting in the analysis of society' (Foucault, 1982, p. 18).

Foucault duly urges historians to find some way around this slipping into the use of total histories – this reliance upon what are plainly modernist metanarratives of one sort or another – and in this connection he proposes the alternative ideal of a 'general history' alert to 'the living, fragile, pulsating history' that so easily 'slips through the fingers' of historians hooked on the modernist obsession with order. It is true that he does not spell out at all systematically what this alternative of general history will entail (perhaps the point is precisely that this alternative cannot be neatly spelled out), but in one highly suggestive passage he writes (1972, p. 10) as follows: 'A total description draws all phenomena around a single centre – a principle, a meaning, a spirit, a world-view, an overall shape: a general history, on the contrary, would deploy the space of a dispersion'. The project of general history is intended as a challenge to the modernist obsession with order, and in respecting the *disorder* of real history beyond the textbooks Foucault insists on paying attention to details as he finds them in the empirical record, as well as on preserving a sense of how the countless phenomena, events, processes and so on found in this record *differ* one from another rather than being basically dimensions of the 'same' central ordering mechanism. Linda Hutcheon (1988, pp. 98–9) therefore writes of Foucault's 'assault on all the centralising forces of unity and continuity', and also of his focus upon 'the particular, the local and the specific' rather than upon 'the general, the universal and the eternal'. And once Foucault's contribution is phrased in this fashion it should immediately be obvious that his criticism of total history and attendant call for a general history is prompted in large measure by a sensitivity to the geography of the human world: a recognition that totalising historical accounts – and, indeed, all modernist metanarratives – necessarily steamroller over the delicate details of how things happen differently in different places, and an attendant recognition that any movement away from these accounts – any movement towards a postmodernist sensibility – should include a much greater awareness of details, difference and geography (see Philo, 1991a, 1991b).

The notion of deploying 'the space of dispersion' that appears in the above quotation from Foucault is very revealing, and this notion has both a metaphorical dimension (Foucault uses it as a metaphor to capture a way of thinking that is alert to difference rather than to sameness) and a methodological dimension (Foucault uses it to emphasise the way in which the simple geography of things 'going on' differently in different places seriously upsets many of the theoretical positions we routinely employ). But the notion also has substantive dimension to it as well, in that built into the middle of Foucault's own historical studies (Foucault, 1967, 1976, 1977, 1979) – which deal with the histories of various social 'pathologies' (madness, sickness, criminality, deviance and abnormal sexuality) and of the institutions invented by society to deal with them (asylums, hospitals, prisons, workhouses and confessionals) – is an attentiveness to tangible 'spaces of dispersion': to the spatial distributions and environmental associations of institutions, to the nearness or farness of institutions from one another and from other phenomena such as human settlements, to the spatial arrangements present in the plans and architectures of these institutions, and to the geographies of the 'discourses' that identify pathologies and then propose institutional solutions to them. As Foucault himself acknowledges, it is a matter of thinking about 'external spaces' and of recognising that these spaces or sites are real, diverse, full of messy things (people, animals, forests, rivers, slopes, buildings, roads, railways: the list is endless) all complexly related to one another:

> The space in which we live, which draws us out of ourselves, in which the erosion [passing] of our lives, our time and our history occurs, the space that claws and knaws at us, is also, in itself, a heterogeneous space. . . . We do not live inside a void . . . , we live inside a set of relations that delineates sites which are irreducible to one another and absolutely not superimposable on one another.
>
> (Foucault, 1986, p. 23)

When Foucault gazes out on the human world of the past, he hence sees not the order of (say) a mode of production determining the lines of class struggle; neither does he see (say) a particular humanly created 'world view' energising everything from how the economy functions to how the most beautiful mural is painted: rather, he immediately sees the 'spaces of dispersion' through which the things under study are scattered across a landscape and related one to another simply through their geography, the only 'order' that is here discernible, by being near to one another or far away; by being positioned in certain places or associated with certain types of environment; by being arranged in a certain way or possessed of a certain appearance thanks to their plans and architectures. It should be added that he does provide an element of more theoretical reflection on the nature of the relations that operate through this geography, and here he discusses how the 'production' of power, knowledge, discourse and 'docile human subjects' are processes that necessarily *take place in space* (see also Dear, 1981; Driver, 1985, 1990; Ogborn, 1990; Philo, 1989a, 1991b), but for present purposes all that needs to be said is that – some inconsistencies in his own studies notwithstanding (Philo, 1991a, 1991b) – what Foucault's work suggests is an intimate connection between the criticism of modernist metanarratives (the total histories) and the conduct of historical inquiries (the general histories) in which a sensitivity to geography is paramount.

We should perhaps say more here about a number of other French theorists (Baudrillard, Deleuze and Derrida) who are identified with a postmodern attitude, and all of whom touch upon the possibilities for a postmodern geography, but the broad sweep of our argument in this section is more or less complete. Lyotard's assessment pivots around an attack on modernist metanarratives for imposing order on a social reality that is thoroughly disordered: a social reality made up of a plurality of different *voices* associated with the ideas, thoughts, hopes, fears and ways of communicating of different peoples in different places. And it is to this range of geographically differentiated voices, this range of 'local knowledges', that a properly conceived postmodern geography must attend. In a sense Foucault then widens the scope of this attack by insisting that modernist metanarratives in the guise of total histories are inevitably insensitive to details, to differences and (in consequence) to *'the differences which make up human geographies'* (Gregory, 1989c, p. 356), and in effectively claiming that any attempt to overcome the limitations of these metanarratives, notably their obsession with order, instantly calls forth an alertness to 'spaces of dispersion' metaphorically, methodologically and more substantively. And it is precisely this alertness to 'spaces of dispersion' in every facet of scholarly inquiry, from inital theoretical reflection through to the rigours of empirical study and writing up of findings (see Figure 6.8 for notes on the task of 'writing'), that is demanded by a properly conceived postmodern geography.

It might be objected that all of this difficult postmodern thinking does nothing more than 'reinvent the wheel': that, in identifying a need to take seriously the *differences* that fracture the human world because it is set in a multiplicity of different spaces, places, environments and landscapes, the postmodern attitude does nothing more than hasten 'an end to geography's alienation from – well, *geography*' (*ibid.* p. 71; and note that Gregory reckons geography to have been 'alienated' from itself by both positivistic and certain post-positivistic approaches). But perhaps this is the whole point, and we therefore feel it unhelpful for geographers to presume that postmodernism is merely another 'approach' to consult for guidelines to structure geographical research: rather, we would suggest that postmodernism is indeed an *attitude* (and not a toolkit of pre-formed concepts and methods), and that it is an attitude that, in itself, insists upon scholarly inquiries being sensitive to the world's geography and thereby perhaps learning from all of the many ways in which the discipline of geography has sought to cultivate and to develop this sensitivity over the years. This is not to imply that geographers cannot greatly improve their self-understanding by

Figure 6.8 'Writing' human geography

Adopting a postmodernist attitude requires us to take far more seriously than hitherto the process of 'writing' human geography: the process of representing our findings in words, sounds and pictures. If we are being critical of all the old certainties (order, truth and so on) – and if we genuinely wish to capture a sense of the myriad differences between people, places and other phenomena of the human world – then we need to take care that the very manner in which we represent our findings does not itself impose too much order or too little awareness of difference upon the complex and even chaotic human reality beyond. We are often urged to write in a highly 'scientific' fashion that stresses the objectivity of the researcher (how often have we been told not to say 'I' in our project reports?) and that implies an isomorphism between the ordered hypothesis-testing of the research and the ordered law-governed nature of the reality beyond, but at no point is it questioned whether this style of writing is in any way appropriate to the subject-matter at hand. And yet it is not just in such 'scientific' writing that the problem arises, since even the more obviously *narrative* presentation – the presentation that tells an 'unfolding story', a chronology of events, a history of causes and effects – risks imposing order and *in*difference upon the subject-matter being addressed. Moreover, this distinctively postmodern 'crisis of representation' is one that strikes home particularly in the context of what Darby (1962; see also Soja, 1986; Gregory, 1989a) once called 'the problem of geographical description', given that writing is inherently *sequential* (we write and read linearly, one word after the next, although it is not clear if we then understand what we are doing in such a purely linear fashion) whilst the human geographies we strive to represent are inherently *simultaneous* (the geography of the human world being all about things 'going on' side by side at the same time).

A geographer such as Olsson (1980, 1987; see also Billinge, 1983) has long sought to question and to extend the languages we employ to represent the human world and, in particular, he has suggested the need to recognise *ambiguity* and to render 'his theoretical reflections *on* discourse . . . themselves deliberate experiments *in* discourse' (Gregory, 1989a, p. 87). Other geographers have experimented with – or have considered what it might mean to experiment with – writing, drawing, painting, photographing, filming and even composing their findings (see, for instance, Balchin, 1972; Jansen, 1984; Daniels, 1985; Soja, 1986; Pocock, 1988b; Quoniam, 1988; Gregory, 1989a), and it is clear that these experiments tie in closely with the new qualitative methodologies – the 'ethnographies' and the 'participant observations' pioneered by anthropologists – that some human geographers are now employing in their efforts to uncover the subtle dynamics of interactions between people and place (Eyles and Smith, 1988: see the relevant sections in Chapter 3). This latter point is actually worth underlining, since a potent source of inspiration for today's human geographers is indeed the debate about researching and representing 'other cultures' that has been prompted among anthropologists such as Geertz (see Geertz, 1973, 1983: and see the papers in Clifford and Marcus, 1984; Marcus and Fischer, 1986). Interestingly enough, leading from a consideration of these anthropological exchanges Gregory (1989a, especially p. 91) concludes that '[t]here *is* a poetics of geography, for geography *is* a kind of writing', and indicates that we need to attend very seriously to the discipline's 'textual strategies' if we are ever to capture the differences, complexities, nuances, achievements and sufferings that *are* the postmodern human geographies of the contemporary world (see also Crang, 1990).

following the writings of Lyotard, Foucault and others, given that these arguments signpost the crucial importance of a geographical sensitivity to the overall debate between modernism and postmodernism, and neither is to suggest that we cannot find in these writings more specific themes (such as those of power and knowledge, particularly 'local knowledge') that could usefully be explored in our own investigations. Yet it is to agree with the pivotal claim of Edward Soja in his insightful recent text, *Postmodern Geographies* (1989, and see Figure 6.9 for notes on this text) that in many ways the postmodern attitude is itself characterised by – is actually 'made by'; is very much 'dependent upon' – a heightened awareness of space: a heightened awareness that the human world in all of its many aspects possesses and is shaped by complex relations of various kinds spun out through an amazing diversity of spaces, places, environments and landscapes (in short, through an amazing diversity of

Figure 6.9 *Edward Soja's* Postmodern Geographies *(1989)*

In this text Soja provides a powerful interpretation of how a sensitivity to *space* has been 'reasserted' in contemporary social theory – he says 'reasserted' because he believes that *modern* social theory (with its roots in *fin de siècle* Vienna) systematically elevated an 'historicism' (a sensitivity to time, chronology and history) over an awareness of space – and in so doing he also makes claims about a range of theoretical developments that amount to both the 'postmodernisation' of geography and the turning of thinkers of all sorts (including Berger, Foucault and Giddens) into 'postmodern geographers'. In this connection Soja can perhaps be criticised for effectively nurturing his own definition of postmodernism, in that he appears to be saying that anybody of whatever disciplinary background who takes space seriously should be designated a 'postmodern' thinker (and hence a 'postmodern geographer'), but we should acknowledge that Soja's arguments here are actually not far removed from our own belief that a postmodern sensibility is necessarily also a geographical one! Up to a point Soja re-states and translates into a fresh vocabulary the claims of his earlier work (notably Soja, 1980; see various passages in Chapter 2), and it is revealing that Lefebvre's arguments about both 'the social production of space' and its 'concealing' function continue to be pivotal to Soja's thinking. Furthermore, Lefebvre is a Marxist urban theorist – albeit one whose 'spatial fetishism' has been criticised on many occasions by other Marxists – and Soja continues to draw heavily upon a version of Marxist 'historical-*geographical* materialism' when turning his attention to what he calls the 'spatiality' of postmodernity. In other words, when he moves away from considering postmodernism as *attitude* to considering the workings of postmodernity as *object* (as the 'geographically uneven development' of the contemporary world; as the 'evocative fragments' of an 'urban region' such as Los Angeles), Soja makes a manoeuvre that really is not so different from that made by Harvey (1989b; see Gregory, 1990a).

human geographies). And perhaps it is appropriate to leave the concluding word in this section to Derek Gregory (1989a, p. 92), whose 'Areal differentiation and postmodern human geography' essay has provided such a useful route-map for this chapter, and who has this to say about the need to consider 'areal differentiation' (an old geographical term – see Hartshorne, 1939 – for which Foucault's 'spaces of a dispersion' is perhaps a (post)modern equivalent):

> we need, in part, to go *back* to the question of areal differentiation: but armed with a new theoretical sensitivity towards the world in which we live and the ways in which we represent it. Whether we focus on 'order' or 'disorder' or on the tension between the two – and no matter how we choose to define these terms – we still have to 'look'. We are still making geography.

Coda to postmodern human geography

In keeping with the common postmodern warning about 'neat' conclusions that impose a somewhat artificial 'closure' upon the argument pursued in an academic text, we will finish this chapter simply with a 'coda' that speculates briefly on the future direction of the engagement between human geography and postmodernism. In the first place, we would expect that numerous human geographers will continue the task signposted by Harvey's *Condition of Postmodernity* (1989b) of unravelling the spatial dimensions to postmodernity as object, and that in the process there will be an ongoing debate about both the extent to which the 'space-economy' of the contemporary world is being 'flexibilised' and the extent to which it is possible to 'read off' matters of culture, politics and ideology from the workings of a transforming 'space-economy'. Moreover, we would suggest that within this broad area of inquiry there will be considerable potential for researchers to examine the geographies of quite specific postmodernist phenomena – notably postmodern architectures, buildings, settlement designs, cultural products and related practices – and in so doing to ask about the uneven diffusion of such phenomena through the cities, towns, villages and countrysides of different regions and nations.

In the second place, though, we expect that human geographers will increasingly come to recognise the gravity of the challenge that postmodernism as attitude poses to most conventional theorisations of the human world, and will begin to appreciate that a sensitivity to the geography of this world – to its fragmentation across multiple spaces, places, environments and landscapes – is itself very much bound up with (and an impetus for) a postmodernist suspicion of modernist 'grand theories' and 'metanarratives'. The arguments to be pursued here will probably take on a variety of guises, although it is likely that Continental theorists will continue to be an influence, alongside an appeal both to Geertzian 'interpretative anthropology' – with its urging of researchers to be humble in not setting their own interpretations of social life above those of the people they study (see Chapter 3, and see also Crang's (1990) call for an ethnographic 'polyphony' allowing many different 'voices' to speak in geographical work on the 'service society') – and to feminist claims about needing to avoid the peculiarly 'masculine' and 'patriarchal' penchant for obliterating difference beneath 'mega-theories' (see Johnson, 1987; and see also Owens, 1985). These manoeuvres will doubtless cause much unease and controversy, however, given that they cast doubt on the stability of the foundations from which most human geography has proceeded over the last thirty or so years (whether these be positivist, Marxist, humanist, structurationist, realist or whatever). And the worries that geographers such as Harvey might air about postmodernism will not simply be academic ones, for it is acutely obvious, as already indicated above, that in attacking all forms of 'totalising' discourse (on the grounds that such discourses obliterate difference) postmodernist thinking risks undermining any project of working towards alternative human and social states. Indeed, it throws into question the whole spirit of 'enlightened modernism' (see above) and thereby risks denying the pursuit of a moral or political agenda designed to make the world a 'better place' for us all to live in. Just such an ambition has been somewhere present in the efforts of virtually all of those individuals who have sought to secure a viable post-positivist human geography, even if the results for Marxists, humanists and others have often ended up being quite different, and this means that debate at the 'critical edges' between those who subscribe to the spirit of modernism and those who do not is destined to be heated and perhaps quite destructive. But we would also express the hope that a measure of creative dialogue can still ensue between geographers whose intellectual and more personal beliefs seemingly head off in divergent directions: in part because the taking seriously of difference is now accepted by the majority of modernist geographers; and in part because there are radical versions of postmodernism in which the 'local knowledges' of particular peoples in particular places are celebrated not just for their own sake, but in the context of a wider inquiry into – and perhaps as a more practical strategy of 'empowering' – the experiences of those groupings (labour, women, ethnic minorities, the unemployed, the sick and so on) who repeatedly 'lose out' in the face of the more material restructurings endemic to the postmodern condition. Maybe it will be possible for human geographers to continue talking to one another, then, and in the process to respect the differences that separate their approaches rather than seeing these differences as insurmountable barriers to insight.

EPILOGUE

Perhaps five or so years ago a book such as this could have been concluded with an acknowledgement of the diversity contained in its pages, but also with an expression of hope in the possible reconciliation of different approaches to human geography under the banner of some over-arching theoretical position such as an historico-geographical materialism or a Giddensian structuration theory grounded in realist philosophy. Indeed, there was a definite mood anticipating the resolution of all of the difficult theoretical problems encountered in 'approaching human geography' – problems to do with determinism and voluntarism, agency and structure, generality and uniqueness, society and space and so on – from the 'highest' philosophical concepts down to the 'lowest' empirical details. Five years later, however, most commentators on theory in human geography are rather more sceptical about the possibilities of achieving such reconciliation and resolution, in part because they have become increasingly aware that the tensions arising between (say) Marxists and humanists are extremely deep-seated (are sedimented in hugely different ways of looking at the human world) and in any case are ultimately rooted in *metaphysical* questions that are fundamentally undecidable. As Trevor Barnes explains in his reply to our inquiries about his 'experiences' of theory in human geography (see Chapter 1), 'In my role as social theorist, Joan Robinson [an economist] gave me the idea of metaphysics. Although ill-defined it raised the notion that all theory was underlaid by metaphysical ideas that themselves could not be justified'. Furthermore, theoretically informed human geographers have become increasingly aware – if a little reluctantly – of the sorts of arguments described here as 'postmodern', which challenge the credentials of those over-arching or 'totalising' (and here we deliberately say 'totalising' rather than 'grand'; see below) theories that profess to offer ontological and epistemological materials satisfying to theoreticians of all different shapes and sizes, and in all possible circumstances (see Chapter 6).

But in recognising that our book cannot end on a high note identifying *the* way forward for theory in human geography, we do not want to sound negative or to suggest that the discipline has taken a backward step. Rather, we want to stress that what is involved now is indeed an acknowledgement of diversity, but an acknowledgement that need not be a straightforward celebration of, a self-indulgent 'wallowing in', the chaos of different and competing approaches: an acknowledgement that can become a forward step away from the naïvety of totalising theories that obscure more than they reveal, towards a recognition of the possibilities for creative and fruitful *dialogue* between different approaches, each capable of opening a distinctive 'window' on the human-geographical reality beyond. Audrey Kobayashi and Suzanne Mackenzie (1989, p. 11) make precisely this point when they claim that 'Apparently common ground, at least in its epistemological terms, is not available just now. What is available . . . is a community of concern, concern to understand, to communicate and to control the process of change'. Kobayashi and Mackenzie enter the qualification about epistemology because they recognise that theoretical manoeuvres such as structuration and realism do at one level allow the combination of rather different (for instance, Marxist *and* humanist) sets of ontological claims, but that in so doing these manoeuvres still cannot provide epistemological equipment enabling the attainment of knowledge about this multi-faceted ontology that will be satisfying to all concerned (and see our discussion of such matters in Chapter 5). But what remains for Kobayashi and Mackenzie is indeed the 'community of concern' in which dialogue can, should and in some measure does take place, even whilst participants in the dialogue appreciate that they are not talking themselves towards a 'final solution' in which they will be persuaded of the 'rightness' of one single theoretical dream.

In the process it may well be that human geographers will accept the *pragmatist* view of 'horses for courses': the view that certain theoretical materials – certain philosophical assumptions and certain methodological packages – will be appropriate for conducting certain tasks and answering certain questions, whereas certain other theories will be more appropriate in certain other circumstances. Thus, it probably is the case that Marxist tools will be most appropriate when inquiring into the interdependent economic development and attendant international division of labour within the contemporary material world, with the rider that the theories employed must be sensitised to space and to capitalism's uneven colonisation of different places. Alternatively, it probably is the case that such tools do not help so much when inquiring into the everyday mental worlds of everyday people, and that humanistic tools will be more appropriate here, with the rider that the theories employed must be prepared to look beyond the transcendental tendencies of much phenomenology and existentialism to the specifics of local meanings, knowledges and languages. We consider the riders introduced here to be vital, since they directly engage with the issue of what human geography can bring to the domain of theory, but we will defer addressing this issue for a moment to consider the charge that the above line of reasoning condemns us to *relativism*.

We have already touched upon the charge of relativism, in that this is the complaint that objects to scepticism about over-arching theoretical ambitions and about the possibility of adjudicating between the relative merits of different and competing intellectual positions. It has to be admitted that the tenor of thinking rehearsed in this 'Epilogue' does now admit the inevitability of a kind of relativism, and does thereby accept a humility that militates against us ever making 'once and for all' statements to the effect that one theory is indisputably 'better' than another theory. Given such an admission many writers respond that it dooms us to a relativism which is entirely hopeless: a relativism that commits us to an 'anything goes' mentality – a situation that is 'just opinion-swapping', to use a phrase from David Livingstone's reply to our questionnaire – and that is troublesome and even downright dangerous because it cannot be used to under-write more practical projects of human and social improvement (in the manner that Marxist theory, for example, underwrites the transition to socialist societies). There is obviously some justification in this claim, it has quickly become the standard Left response to postmodernism as attitude (see Chapters 2 and 6), but we believe that the situation is actually rather more complicated, and is more complicated for several reasons. In the first place, for instance, the pragmatist view just outlined can be cast in a stronger vein than simply saying that one particular variety of theory seems to 'fit' neatly with one particular segment of social reality under study: thus, we would suggest that it is not just 'appropriate' but *necessary* to draw upon certain theoretical materials – even those that continue to have 'grand' ambitions in seeking to tackle 'large' social issues and 'extensive' geographical areas (such as Marxist theories of under-development) – when confronting certain tasks and certain questions. In other words, whilst maintaining some appreciation of the limits that render no one theory capable of doing everything, we would still argue that the connections between theories and 'things' studied are and should be considerably more than just casual or accidental. In the second place, we are convinced that recognising the limits to theory may actually force academics to confront themselves, their intellectual, moral and political standpoints much more explicitly than is the case where theory – and particularly theory that goes beyond being 'grand' to being 'totalising' – provides easy 'crutches' for leaning on. Indeed, the problem with such arrogant theories is that they (as it were) do much of the hard theoretical, moral and political thinking for the academic without he or she having to input all that much of themselves to the particular project in hand. Once the 'security blanket' of theory has been pulled away, it is hence no longer possible for individuals simply to appeal to the theoretical authority of great thinkers such as Durkheim, Marx or Weber, and neither is it possible for them to hide their moral and political choices behind the supposed objectivity and 'scientism' of the research process. In short, it may be that acknowledging the relativism of our studies as human geographers actually forces us to be more explicit about our own theorising and to be more honest about the moral and political bases of the choices we make about both what to study and how to study.

But, having reached this point, we would also want to admit that for many of us – whilst perhaps appreciating the 'philosophical' logic of accepting a measure of relativism – at bottom what drives our own 'geographical imag-ination' is something that continues to be energised by quite fundamental beliefs, which cannot readily accept their 'equality' with other sets of beliefs. For many of us, then, there is still a desire to cling on to profound and seemingly consistent frames of reference for the practice of our human geo-graphy, and it is surely the case that nobody has the right to deny the legit-imacy of this desire (even if they may wish to be critical of the particular frames of reference alighted upon). These frames of reference are themselves quite varied, of course, and we are attracted to particular frames for different 'reasons': sometimes in a manner we can reflect upon and rationalise, but often in a manner that is more unconscious, taken-for-granted and probably bound up with our individual education and socialisation. In part, these frames of reference are *intellectual-theoretical* ones, since many of us have a fiercely held academic commitment to the philosophy and methods of conventional 'science', of Marxism, of humanism or of other approaches to knowledge, and will duly argue at length the pros and cons of this intellectual-theoretical commitment.

In part, though, the most crucial frames of reference we are discussing here are rooted 'beyond' academic matters – whilst on occasion still informing just what it is that we do in the academy – and it is, it is hoped, apparent that we are now addressing issues to do with the geographer's *morality* and *politics*. Perhaps most obviously on this terrain we encounter a deeply held commitment to a radical and transformative politics capable of changing the world, and in this respect it is clear that many geographers have a keen sense of the rights and wrongs associated with particular political stances, and a concomitant wish to inform political practice through their academic geographical work. Inevitably, this 'mix' of politics and geography generates a particular way of looking at the world that suggests the questions to be asked, how to ask them and what to do with the answers to them, and the most obvious example here is given in the closely interlinked politics and geography of Marxists such as David Harvey. Similarly, another strand of attachment to fundamentals found among geographers stems from a strong religious faith, and it is evident that the Christian geographer (for example) will bring to his or her work a specific value-system and a particular portfolio of 'truths' that are felt to be God-given, and that are hence assumed to be 'more true' than any opposing claims made by non-Christian geographers. This individual will subscribe to fundamental views of right and wrong (and of good and bad) in much the same way as will the Marxist geographer, and it may be that these two sets of views will overlap in a meaningful and mutually informing fashion (although this certainly need not be the case). It is not difficult to see that reconciling such deeply held convictions – with a relativism in approaching human geography is a difficult matter, and it is hence no wonder that the debate about relativism in geography (and about the apparent 'creep' towards relativism traced through the chapters of this book) is a tense, heated but also somewhat confused one. All that we can offer in this 'Epilogue' is a preference for a kind of relativism that accepts the incompatibility of different approaches to human geography, but that respects this incompatibility – recognising the genuine depth of attachment to different intellectual, political, moral, ethical and faith-centred fundamentals that it expresses – and that still strives to allow incompatibles to enter into constructive (rather than destructive) dialogue.

These are difficult and discomfiting arguments, it must be said, but equally difficult (if not so unsettling to us as human geographers) are the arguments that we will now conclude with here, and that pivot around the suggestion that human geography is now itself saying something back to philosophy and social theory, as opposed simply to picking up bits and pieces of philosophy and social theory that can inform human-geographical inquiries. The respondents to our questionnaire actually came up with somewhat divergent opinions in this respect, in that a few feel that human geography still has nothing of any theoretical consequence to say back to philosophy and social theory whereas others feel that it does now have something to say. Here are two of the more negative assessments:

> There is actually very little original theoretical work in geography, and I personally look forward to geographers producing more and better theoretically informed monographs.
>
> (Gerry Kearns)

> I tend to be very cynical about geographical theory (or, rather, theoretical ideas put forward by geographers). . . . Geographers are appalling theoreticians when it comes to process – structure and spatial pattern have been the centre of attention – speculation and assertion is what is offered on dynamism/process. This is at the core of my disquiet.
>
> (Keith Hoggart)

But some of the assessments were more positive:

> Geographers are borrowing from other social scientists – sociologists, anthropologists, political scientists and others – and are feeling more confident in saying that geography matters as well. . . . Perhaps the most exciting thing about new theoretical approaches is the opportunity to show specific ways in which social, cultural, etc., theories complement geographical concepts (e.g. ghettoisation of the mentally ill has spatial as well as cultural, social and political aspects).
>
> (Wil Gesler)

> Still, despite the usual unseemly scramble for novelty, the Niagaras of name-dropping and the weakly founded self-confidence, a great deal of progress was made. 'Geographers' caught up and began to make an impression. The discipline survived as a congenial flag of convenience, and the subject moved a step or two up the academic status ladder.
>
> (Andrew Sayer)

We would tend to side with these more positive interpretations, whilst wishing to echo the careful nature of their claims (which are not so self-assured as those contained in a text such as Soja, 1989), and in so doing would identify at least two ways in which a sensitivity to the human geography of the world (if not necessarily to a discipline called 'human geography') is indispensable to contemporary philosophy and social theory. In the first instance, a sensitivity to geography – to the 'areal differentiation' of people, events and phenomena simply being different in different places – must always be kept in mind as a counter to the 'totalising' ambitions of 'grand theories' that reckon the world 'to happen' in a curiously spaceless head-of-a-pin realm. And this is what Andrew Kirby perhaps means in his questionnaire response where he suggests that 'geography is a deconstructive lens', and this is undoubtedly what David Harvey meant several years earlier when he insisted that 'the insertion of concepts of space, place, locale and milieu into any social theory has a numbing effect on that theory's central propositions' (1984, p. 8). In the second instance, and as argued in various ways throughout this book, the human world has a geography (in a very real sense, it is a geography), and so to inquire theoretically into the workings of economy, society, politics and culture is inevitably to inquire about their geographies: it is to ask about where 'things' are and why, and about how 'things' are connected up one to another in place and across space. These twin aspects of how geography 'speaks' to theorisations of the human world need themselves to be theorised, of course, and this means that we need to be able to talk theoretically about both the difference human geography makes to intellectual activity and the differences that are so crucially present in the geography of the human world. A great many human geographers are hence striving in one way or another to develop new forms of 'geographic theory' capable of tackling and reflecting these twin aspects, and Robert Lake sums up this endeavour in a manner that neatly captures both the sweep and the spirit of our book:

Theoretical sophistication has replaced quantitative/analytical sophistication as the guarantor of geography's legitimacy. This in a way is much more problematic for the discipline. Methodological rigour is after all easier to demonstrate than is theoretical rigour. But I'm convinced that by our theories they will know us. The irony is that there's still a long, long way to go. The quantitative revolution made us analytical and legitimised abstraction (within the discipline). In this sense, it was a necessary precursor. Now that we have that foundation, we need to [do] a great deal of work in developing the substance of geographical theory. We're just beginning.

REFERENCES

Abler, R. (1987) What shall we say? To whom shall we speak?, *Annals of the Association of American Geographers*, Vol. 77, pp. 511–24.

Ackerman, E. A. (1945) Geographic training, wartime research and immediate professional objectives, *Annals of the Association of American Geographers*, Vol. 35, pp. 121–43.

Acton, H. B. (1967) Idealism, entry in P. Edwards (ed.) *The Encyclopaedia of Philosophy, Volume 4*, Collier-Macmillan, London, pp. 110–18.

Aglietta, M. (1979) *A Theory of Capitalist Regulation*, New Left Books, London.

Albertsen, N. (1988) Postmodernism, post-Fordism and critical social theory, *Environment and Planning D: Society and Space*, Vol. 6, pp. 339–65.

Allen, J. (1983) Property relations and landlordism: a realist approach, *Environment and Planning D: Society and Space*, Vol. 1, pp. 191–203.

Althusser, L. (1969) *For Marx*, Penguin Books, Harmondsworth.

Althusser, L. and Balibar, E. (1970) *Reading Capital*, New Left Books, London.

Amin, A. (1989) Flexible specialization and small firms in Italy: myths and realities, *Antipode*, Vol. 21, pp. 13–34.

Amin, S. (1976) *Unequal Development*, Monthly Review Press, New York, NY.

Amin, S. (1980) *Class and Nation: Historically and in the Current Crisis*, Heinemann, London.

Anderson, J. (1973) Ideology in geography: an introduction, *Antipode*, Vol. 5, no. 3, pp. 1–6 (reprinted in *Antipode*, 1985, op. cit.).

Antipode (1974) B. Berry's review of D. Harvey (1973) *Social Justice and the City*, reply by D. Harvey and debate, Vol. 6, no. 2, pp. 142–9.

Antipode (1975) Union of socialist geographers: seminar in Marxist geography, *Antipode*, Vol. 5, no. 3, pp. 54–85.

Antipode (1985) The best of *Antipode*, 1969–1985, *Antipode*, Vol. 17, nos. 2 & 3.

Archer, K. (1987) Mythology and the problem of reading in urban and regional research, *Environment and Planning D: Society and Space*, Vol. 5, pp. 384–93.

Ashley, D. (1982) Historical materialism and social evolution, *Theory, Culture and Society*, Vol. 1, pp. 89–92.

Bailey, P. (1979) 'Will the real Bill Banks please stand up?' Towards a role analysis of mid-Victorian working-class respectability, *Journal of Social History*, Vol. 12, pp. 336–52.

Balchin, W. G. V. (1972) Graphicacy, *Geography*, Vol. 57, pp. 185–95.

Barnes, T. and Curry, M. (1983) Towards a contextualist approach to geographical knowledge, *Transactions of the Institute of British Geographers* (NS), Vol. 8, pp. 467–82.

Bell, D. (1973) *The Coming of Post-Industrial Society*, Basic Books, New York, NY.

Bennett, R. J. (1985) Quantification and relevance, in R. J. Johnston (ed.), op. cit.

Benton, T. (1977) *Philosophical Foundations of the Three Sociologies*, Routledge & Kegan Paul, London.

Benton, T. (1981) Realism and social science, *Radical Philosophy*, no. 27, pp. 13–21.

Berdoulay, V. (1976) French possibilism as a form of neo-Kantian philosophy, *Proceedings of the Association of American Geographers*, Vol. 8, pp. 176–9.

Berdoulay, V. (1978) The Vidal–Durkheim debate, in D. Ley and M. S. Samuels (eds.), op. cit.

Berger, P. and Luckmann, T. (1967) *The Social Construction of Reality: A Treatise in the Sociology of Knowledge*, Allen Lane, London.

Berman, M. (1982) *All That Is Solid Melts Into Air: The Experience of Modernity*, Verso, London.

Berry, B. J. L. (1970) The geography of the United States in the year 2000, *Transactions of the Institute of British Geographers*, Vol. 51, pp. 21–53.

Berry, B. J. L. (1972a) More on relevance and policy analysis, *Area*, Vol. 4, pp. 77–80.

Berry, B. J. L. (1972b) Revolutionary and counter-revolutionary theory in geography: a ghetto commentary, *Antipode*, Vol.4, no.2, pp. 31–3.

Beynon, H. and Hudson, R. with Sadler, D. (1991) *A Tale of Two Industries: The Contraction of Coal and Steel in the North East of England*, Open University Press, Milton Keynes.

Bhaskar, R. (1975) *A Realist Theory of Science*, Leeds Books, Leeds (reprinted in 1978 by Harvester, Brighton).

Bhaskar, R. (1979) *The Possibility of Naturalism: A Philosophical Critique of the Contemporary Human Sciences*, Harvester, Brighton.

Bhaskar, R. (1980) Scientific explanation and human emancipation, *Radical Philosophy*, no. 26, pp. 16–28.

Bhaskar, R. (1982) Emergence, explanation and emancipation, in P. Secord (ed.) *Explaining Human Behavior*, Sage, Beverly Hills, Calif.

Bhaskar, R. (1986) *Scientific Realism and Human Emancipation*, Verso, London.

Bhaskar, R. (1989) *Reclaiming Reality: A Critical Introduction to Contemporary Philosophy*, Verso, London.

Billinge, M. (1977) In search of negativism: phenomenology and historical geography, *Journal of Historical Geography*, Vol. 3, pp. 55–68.

Billinge, M. (1983) The Mandarin dialect: an essay on style in contemporary geographical writing, *Transactions of the Institute of British Geographers* (NS), Vol. 8, pp. 400–20.

Blaikie, P. (1978) The theory of the spatial diffusion of innovations: a spacious cul-de-sac, *Progress in Human Geography*, Vol. 2, pp. 268–95.

Blaut, J. M. (1970) Geographic models of imperialism, *Antipode*, Vol. 2, no. 1, pp. 65–85 (reprinted in *Antipode*, 1985, op. cit.).

Blaut, J. M. (1973) The theory of development, *Antipode*, Vol. 5, no. 2, pp. 22–6 (reprinted in R. Peet (ed.) (1978), op. cit.).

Blaut, J. M. (1975) Imperialism: the Marxist theory and its evolution, *Antipode*, Vol. 7, no. 1, pp. 1–19.

Blaut, J. M. (1984) Modesty and the movement: a commentary, in T. F. Saarinen, D. Seamon and J. L. Sell (eds.), op. cit.

Bleicher, J. and Featherstone, M. (1982) Historical materialism today: an interview with Anthony Giddens, *Theory, Culture and Society*, Vol. 1, pp. 63–77.

Blowers, A. T. (1972) Bleeding hearts and open values, *Area*, Vol. 4, pp. 290–2.

Boddy, M. (1976) The structure of mortgage finance: building societies and the British social formation, *Transactions of the Institute of British Geographers* (NS), Vol. 1, pp. 58–71.

Bottomore, T. (ed.) (1973) *Karl Marx*, Blackwell, Oxford.

Bottomore, T. (ed.) (1981) *Modern Interpretations of Marx*, Blackwell, Oxford.

Bourdieu, P. (1977) *Outline of a Theory of Practice*, Cambridge University Press.

Bowen, E. G. (1976) Herbert John Fleure and Western European geography in the twentieth century, reprinted in E. G. Bowen, *Geography, Culture and Habitat: Selected Essays (1925–1975) of E. G. Bowen*, Gomer Press, Llandysul.

Bowen, M. (1981) *Empiricism and Geographical Thought: From Francis Bacon to Alexander von Humboldt*, Cambridge University Press.

Bradbury, M. (1988) Modernism and Post-modernism, entries in A. Bullock, O. Stallybrass and S. Trombley (eds.) *The Fontana Dictionary of Modern Thought*, Fontana, London, pp. 539–40 and pp. 671–2.

Breitbart, M. (1978/1979) Anarchist decentralization in rural Spain, 1936–1939: the integration of community and environment, *Antipode*, Vol. 10, no. 3, Vol. 11, no. 1, pp. 83–98 (reprinted in *Antipode*, 1985, op. cit.).

Breitbart, M. (1981) Peter Kropotkin, the anarchist geographer, in D. R. Stoddart (ed.) *Geography, Ideology and Social Concern*, Blackwell, Oxford.

Brusco, S. (1982) The Emilian model: productive decentralisation and social integration, *Cambridge Journal of Economics*, Vol. 6, pp. 167–84.

Buber, M. (1957) Distance and relation, *Psychiatry*, Vol. 201, pp. 97–104.

Buchanan, K. (1974) Reflections on a 'dirty word', *Dissent*, no.31, pp. 25–31 (reprinted in R. Peet (ed.) (1978), op. cit.).

Bunge, W. (1962) Theoretical geography (Lund Studies in Geography, Series C: General and Mathematical Geography, Paper no. 1), Gleerup, Lund.

Bunge, W. (1969) The first years of the Detroit Geographical Expedition: a personal report, *Field Notes*, no. 1, pp. 1–9 (reprinted in R. Peet (ed.), 1978, op. cit. Page nos. in text refer to this reprint).

Bunge, W. (1971) *Fitzgerald: The Geography of a Revolution*, Schenkman, Cambridge, Mass.

Bunge, W. (1973) The geography of human survival, *Annals of the Association of American Geographers*, Vol. 63, pp. 275–95.

Bunting, T. E. and Guelke, L. (1979) Behavioral and perception geography: a critical appraisal, *Annals of the Association of American Geographers*, Vol. 69, pp. 448–62.

Burgess, J. (1988) Group analysis and field research in Greenwich, *Social and Cultural Geography Study Group Newsletter*, Spring, pp. 5–8.

Burton, I. (1963) The quantitative revolution and theoretical geography, *Canadian Geographer*, Vol. 7, pp. 151–62.

Buttimer, A. (1971) *Society and Milieu in the French Geographic Tradition*, Rand McNally, Chicago, Ill.

Buttimer, A. (1974) Values in geography (Association of American Geographers, Commission on College Geography, Resource Paper no. 24), Washington, DC.

Buttimer, A. (1976) Grasping the dynamism of the life-world, *Association of American Geographers*, Vol. 66, pp. 277–92.

Buttimer, A. (1978) Charism and context: the challenge of *la geographie humaine*, in D. Ley and M. S. Samuels (eds.), op. cit.

Buttimer, A. (1979) Reason, rationality and human creativity, *Geografiska Annaler*, Vol. 61B, pp. 43–9.

Buttimer, A. (1980) Home, reach and a sense of place, in A. Buttimer and D. Seamon (eds.), op. cit.

Buttimer, A. (ed.) (1983) *The Practice of Geography*, Longman, London.

Buttimer, A. and Seamon, D. (eds.) (1980) *The Human Experience of Space and Place*, Croom Helm, London.

Callinicos, A. (1989) Anthony Giddens: a contemporary critique, in A. Callinicos (ed.) *Marxist Theory*, Oxford University Press.

Campbell, K. (1967) Materialism, entry in P. Edwards (ed.) *The Encyclopaedia of Philosophy, Volume 5*, Collier-Macmillan, London, pp. 179–88.

Carlstein, T. (1981) The sociology of structuration in time and space: a time-geographic assessment of Giddens's theory, *Svensk Geografisk Arsbok*, no. 57, pp. 41–57.

Carney, J. and Hudson, R. with Lewis, J. (1976) Regional under-development in late capitalism: a study of the North East of England, in I. Masser (ed.) *Theory and Practice in Regional Science*, Pion, London.

Carney, J. and Hudson, R. with Lewis, J. (eds.) (1980) *Regions in Crisis: New Perspectives in European Regional Theory*, Croom Helm, London.

Castells, M. (1977) *The Urban Question*, Edward Arnold, London.

Chappell, J. E. (1981) The phenomenological environmentalism of Watsuji Tetsuro, *History of Geography Newsletter*, Vol. 1, pp. 7–14.

Chatwin, B. (1987) *The Songlines*, Jonathan Cape, London.

Chorley, R. J. and Haggett, P. (eds.) (1965) *Frontiers in Geographical Teaching*, Methuen, London.

Chorley, R. J. and Haggett, P. (eds.) (1967) *Models in Geography*, Methuen, London.

Chouinard, V. and Fincher, R. (1983) A critique of 'Structural Marxism and human geography', *Annals of the Association of American Geographers*, Vol. 73, pp. 137–46.

Christensen, K. (1982) Geography as a human science: a philosophic critique of the positivist-humanist split, in P. Gould and G. Olsson (eds.) *A Search for Common Ground*, Pion, London.

Clark, J. and Modgil, C. with Modgil, S. (eds.) (1990) *Anthony Giddens: Consensus and Controversy*, Famer Press, Lewes.

Clark, M. J. and Gregory, K. J. with Gurnell, A. M. (eds.) (1989) *Horizons in Physical Geography*, Macmillan, London.

Clifford, J. and Marcus, G. E. (eds.) (1984) *Writing Culture: The Poetics and Politics of Ethnography*, University of California Press, Berkeley, Calif.

Cloke, P. (1987) Rurality and change: some cautionary notes, *Journal of Rural Studies*, Vol. 3, pp. 71–6.

Cloke, P. (1989) Rural geography and political economy, in R. Peet and N. Thrift (eds.), op. cit., Vol. I.

Cochrane, A. (1987) What a difference the place makes: the new structuralism of locality, *Antipode*, Vol. 19, pp. 354–63.

Cohen, I. (1989) *Structuration Theory*, Macmillan, London.

Cooke, P. (ed.) (1986a) *Global Restructuring, Local Response*, ESRC, London.

Cooke, P. (1986b) The changing urban and regional system in the United Kingdom, *Regional Studies*, Vol. 20, pp. 243–51.

Cooke, P. (1987a) Britain's new spatial paradigm: technology, locality and society in transition, *Environment and Planning A*, Vol. 19, pp. 1289–301.

Cooke, P. (1987b) Clinical inference and geographic theory, *Antipode*, Vol. 19, pp. 407–16.

Cooke, P. (ed.) (1989a) *Localities: The Changing Face of Urban Britain*, Unwin Hyman, London.

Cooke, P. (1989b) Locality theory and the poverty of 'spatial variation', *Antipode*, Vol. 21, pp. 261–73.

Cooke, P. (1989c) The contested terrain of locality studies, *Tijdschrift voor Economische en Sociale Geografie*, Vol. 80, pp. 14–29.

Coppock, J. T. (1974) Geography and public policy: challenges, opportunities and implications, *Transactions of the Institute of British Geographers*, Vol. 63, pp. 1–16.

Cosgrove, D. (1979) John Ruskin and the geographical imagination, *Geographical Review*, Vol. LXIX, pp. 43–62.

Cosgrove, D. (ed.) (1982) Geography and the humanities (Loughborough University of Technology, Department of Geography, Occasional Paper no. 5), Loughborough.

Cosgrove, D. (1984) *Social Formation and Symbolic Landscape*, Croom Helm, London.

Cosgrove, D. (1985) Prospect, perspective and the evolution of the landscape idea, *Transactions of the Institute of British Geographers* (NS), Vol. 10, pp. 43–62.

Cosgrove, D. (1989a) Geography is everywhere: culture and symbolism in human landscapes, in D. Gregory and R. Walford (eds.), op. cit.

Cosgrove, D. (1989b) Historical considerations on humanism, historical materialism and geography, in A. Kobayashi and S. Mackenzie (eds.), op. cit.

Cosgrove, D. (1989c) Models, description and imagination in geography, in B. Macmillan (ed.), op. cit.

Cosgrove, D. and Daniels, S. (eds.) (1988) *The Iconography of Landscape: Essays on the Symbolic Representation, Design and Use of Past Environments*, Cambridge University Press.

Cosgrove, D. and Jackson, P. (1987) New directions in cultural geography, *Area*, Vol. 19, pp. 95–101.

Couclelis, H. and Golledge, R. G. (1983) Analytic research, positivism and behavioral geography, *Annals of the Association of American Geographers*, Vol. 73, pp. 331–9.

Cox, K. and Mair, A. (1989) Levels of abstraction in locality studies, *Antipode*, Vol. 21, pp. 121–32.

Cox, N. J. (1989) Modelling, data analysis and Pygmalion's problem, in B. Macmillan (ed.), op. cit.

Crang, P. (1990) Contrasting images of the new service society, *Area*, Vol. 22, pp. 29–36.

Crilley, D. (1990) The disorder of John Short's new urban order, *Transactions of the Institute of British Geographers* (NS), Vol. 15, pp. 232–8.

Cullen, I. G. (1976) Human geography, regional science and the study of individual behaviour, *Environment and Planning A*, Vol. 8, pp. 397–410.

Cullen, J. and Knox, P. (1982) The city, the self and urban society, *Transactions of the Institute of British Geographers* (NS), Vol. 7, pp. 276–91.

Curry, M. (1982a) The idealist dispute in Anglo-American geography, *Canadian Geographer*, Vol. 24, pp. 37–50.

Curry, M. (1982b) The idealist dispute in Anglo-American geography: a reply to Guelke, *Canadian Geographer*, Vol. 24, pp. 57–9.

Dallmayr, F. (1982) The theory of structuration: a critique, in A. Giddens, *Profiles and Critiques in Social Theory*, Macmillan, London.

Daniels, S. (1985) Arguments for a humanistic geography, in R. J. Johnston (ed.), op. cit.

Daniels, S. (1989) Marxism, culture and the duplicity of landscape, in R. Peet and N. Thrift (eds.), op. cit., Vol. II.

Darby, H. C. (1962) The problem of geographical description, *Transactions of the Institute of British Geographers*, Vol. 30, pp. 1–14.

Davis, W. M. (1915) The principles of geographical description, *Annals of the Association of American Geographers*, Vol. 5, pp. 61–105.

Dawe, A. (1970) The two sociologies, *British Journal of Sociology*, Vol. 21, pp. 207–18.

Dawe, A. (1979) Theories of social action, in T. Bottomore and R. Nisbet (eds.) *A History of Sociological Analysis*, Heinemann, London.

Dear, M. (1975) The nature of socialist geography, *Antipode*, Vol. 7, no. 1, pp. 87–9.

Dear, M. (1981) Social and spatial reproduction of the mentally ill, in M. Dear and A. J. Scott (eds.) *Urbanisation and Urban Planning in Capitalist Society*, Methuen, London.

Dear, M. (1986) Postmodernism and planning, *Environment and Planning D: Society and Space*, Vol. 4, pp. 367–84.

Dear, M. (1988) The postmodern challenge: reconstructing human geography, *Transactions of the Institute of British Geographers* (NS), Vol. 13, pp. 262–74.

Dear, M. J. and Moos, A. I. (1986) Structuration theory in urban analysis: 2. empirical application, *Environment and Planning A*, Vol. 18, pp. 351–73.

Derrida, J. (1967) *De la Grammatologie*, Les Editions de Minuit, Paris (translated as J. Derrida (1974) *Of Grammatology*, Johns Hopkins University Press, Baltimore Md.).

Dickie-Clark, H. F. (1984) Anthony Giddens's theory of structuration, *Canadian Journal of Political Social Theory*, Vol. 8, pp. 92–110.

Dodgshon, R. (1987) *The European Past: Social Evolution and Spatial Order*, Macmillan, London.

Driver, F. (1985) Power, space and the body: a critical assessment of Foucault's *Discipline and Punish, Environment and Planning D: Society and Space*, Vol. 3, pp. 425–46.

Driver, F. (1990) Discipline without frontiers? Representations of the Mettray Reformatory Colony in Britain, 1840–1880, *Journal of Historical Sociology*, Vol. 3, pp. 272–93.

Duncan, J. (1978) The social construction of unreality: an interactionist approach to the tourist's cognition of environment, in D. Ley and M. S. Samuels (eds.), op. cit.

Duncan, J. (1985) Individual action and political power: a structuration perspective, in R. J. Johnston (ed.), op. cit.

Duncan, J. and Ley, D. (1982) Structural Marxism and human geography: a critical perspective, *Annals of the Association of American Geographers*, Vol. 72, pp. 30–59.

Duncan, S. (1989a) Uneven development and the difference that space makes, *Geoforum*, Vol. 20, pp. 131–40.

Duncan, S. (1989b) What is locality?, in R. Peet and N. Thrift (eds.), op. cit., Vol. II.

Duncan, S. and Goodwin, M. (1988) *The Local State and Uneven Development*, Polity Press, Cambridge.

Duncan, S. and Savage, M. (1989) Space, scale and locality, *Antipode*, Vol. 21, pp. 179–206.

Eco, U. (1983) *The Name of the Rose*, Secker & Warburg, London.

Edwards, D. L. (1988) Enlightenment, entry in A. Bullock, O. Stallybrass and S. Trombley (eds.) *The Fontana Dictionary of Modern Thought*, Fontana, London, pp. 272–3.

Eliot Hurst, M. E. (1973) Establishment geography, *Antipode*, Vol. 5, no. 2, pp. 40–59.

Eliot Hurst, M. E. (1980) Geography, social science and society: towards a de-definition, *Australian Geographical Studies*, Vol. 18, pp. 3–21.

Eliot Hurst, M. E. (1985) Geography has neither existence nor future, in R. J. Johnston (ed.), op. cit.

Entrikin, J. N. (1976) Contemporary humanism in geography, *Annals of the Association of American Geographers*, Vol. 66, pp. 615–32.

Entrikin, J. N. (1977) Geography's spatial perspective and the philosophy of Ernst Cassirer, *Canadian Geographer*, Vol. 21, pp. 209–22.

Entrikin, J. N. (1981) Royce's 'provincialism': a metaphysician's social geography, in D. R. Stoddart (ed.) *Geography, Ideology and Social Concern*, Blackwell, Oxford.

Evans, D. M. (1978) Alienation, mental illness and the partitioning of space, *Antipode*, Vol. 10, no. 1, pp. 13–23.

Eyles, J. (1981) Why geography cannot be Marxist: towards an understanding of lived experience, *Environment and Planning A*, Vol. 12, pp. 1371–88.

Eyles, J. (1989) The geography of everyday life, in D. Gregory and R. Walford (eds.), op. cit.

Eyles, J. and Smith, D. M. (eds.) (1988) *Qualitative Methods in Human Geography*, Polity Press, Cambridge.

Farber, M. (1960) Phenomenology, entry in J. O. Urmson (ed.) *The Concise Encyclopaedia of Western Philosophy and Philosophers*, Hutchinson, London, pp. 292–5.

Febvre, L. (1925) *A Geographical Introduction to History*, Alfred Knopf, New York, NY.

Fell, J. P. (1979) *Heidegger and Sartre: An Essay on Being and Place*, Columbia University Press, New York, NY.

Findlay, J. N. (1960) Husserl, entry in J. O. Urmson (ed.) *The Concise Encyclopaedia of Western Philosophy and Philosophers*, Hutchinson, London, pp. 188–90.

Fine, B. (1975) *Marx's Capital*, Macmillan, London.

Fine, B. and Harris, L. (1979) *Rereading Capital*, Macmillan, London.

Fleure, H. J. (1965) Recollections of A. J. Herbertson, *Geography*, Vol. L, pp. 348–9.

Folke, S. (1972) Why a radical geography must be Marxist, *Antipode*, Vol. 4, no. 2, pp. 13–18.

Foord, J. and Gregson, N. (1986) Patriarchy: towards a reconceptualization, *Antipode*, Vol. 18, pp. 186–211.

Forbes, D. (1984) *The Geography of Underdevelopment*, Croom Helm, London.

Foucault, M. (1967) *Madness and Civilization: A History of Insanity in the Age of Reason*, Tavistock, London.

Foucault, M. (1972) *The Archaeology of Knowledge*, Tavistock, London.

Foucault, M. (1976) *The Birth of the Clinic: An Archaeology of Medical Perception*, Tavistock, London.

Foucault, M. (1977) *Discipline and Punish: The Birth of the Prison*, Allen Lane, London.

Foucault, M. (1979) *The History of Sexuality, Volume 1*, Allen Lane, London.

Foucault, M. (1982) Interview with Michel Foucault on space, knowledge and power, *Skyline*, March, pp. 17–20.

Foucault, M. (1986) Of other spaces, *Diacritics*, Spring, pp. 22–7.

Frank, A. G. (1971) *Capitalism and Underdevelopment*, Penguin Books, Harmondsworth.

Frazier, J. W. (1981) Pragmatism: geography and the real world, in M. E. Harvey and B. P. Holly (eds.), op. cit.

Gadamer, H. G. (1975) *Truth and Method*, Sheed & Ward, London.

Galois, B. (1976) Ideology and the idea of nature: the case of Peter Kropotkin, *Antipode*, Vol. 8, no. 3, pp. 1–16 (reprinted in R. Peet (ed.), 1978, op. cit.).

Gane, M. (1983) Anthony Giddens and the crisis of social theory, *Economy and Society*, Vol. 12, pp. 368–98.

Gay, P. (1969) *The Enlightenment: An Interpretation – Volume II: The Science of Freedom*, Weidenfeld & Nicolson, London.

Geertz, C. (1973) *The Interpretation of Cultures: Selected Essays*, Basic Books, New York, NY.

Geertz, C. (1983) *Local Knowledge: Further Essays in Interpretative Anthropology*, Basic Books, New York, NY.

Gibson, E. M. W. (1981) Realism, in M. E. Harvey and B. P. Holly (eds.), op. cit.

Giddens, A. (1971) *Capitalism and Modern Social Theory*, Cambridge University Press.

Giddens, A. (1976) *New Rules of Sociological Method*, Hutchinson, London.

Giddens, A. (1977) *Studies in Social and Political Theory*, Hutchinson, London.

Giddens, A. (1979) *Central Problems in Social Theory: Action, Structure and Contradiction in Social Analysis*, Macmillan, London.

Giddens, A. (1981) *A Contemporary Critique of Historical Materialism, Volume 1: Power, Property and the State*, Macmillan, London.

Giddens, A. (1982) A reply to my critics, *Theory, Culture and Society*, Vol. 1, pp. 107–13.

Giddens, A. (1984) *The Constitution of Society: Outline of the Theory of Structuration*, Polity Press, Cambridge.

Giddens, A. (1985a) *A Contemporary Critique of Historical Materialism, Volume 2: The Nation-State and Violence*, Polity Press, Cambridge.

Giddens, A. (1985b) Time, space and regionalisation, in D. Gregory and J. Urry (eds.), op. cit.

Giddens, A. (1989a) *A Contemporary Critique of Historical Materialism, Volume 3: Between Capitalism and Socialism*, Polity Press, Cambridge.

Giddens, A. (1989b) A response to my critics, in D. Held and J. B. Thompson (eds.), op. cit.

Glacken, C. J. (1967) *Traces on the Rhodian Shore: Nature and Culture in Western Thought from Ancient Times to the End of the Eighteenth Century*, University of California Press, Berkeley, Calif.

Gold, J. R. (1980) *An Introduction to Behavioural Geography*, Oxford University Press.

Golledge, R. G. and Amadeo, D. (1968) On laws in geography, *Annals of the Association of American Geographers*, Vol. 58, pp. 760–74.

Golledge, R. G. and Couclelis, H. (1984) Positivist philosophy and research on human spatial behavior, in T. F. Saarinen, D. Seamon and J. L. Sell (eds.), op. cit.

Golledge, R. G. and Rushton, G. (eds.) (1976) *Spatial Choice and Spatial Behavior: Geographic Essays on the Analysis of Preferences and Perceptions*, Ohio State University Press, Columbus, OH.

Goss, J. (1988) The built environment and social theory, *Professional Geographer*, Vol. 40, pp. 392–403.

Gottdiener, M. (1987) Space as a factor of production, *International Journal of Urban and Regional Research*, Vol. 11, pp. 405–16.

Gould, P. (1970) Is 'statistix inferens' the geographical name for a wild goose?, *Economic Geography*, Vol. 46, pp. 439–48.

Gould, P. (1976) Cultivating the garden: a commentary and critique on some multidimensional speculations, in R. G. Golledge and G. Rushton (eds.), op. cit.

Gould, P. and White, R. (1974) *Mental Maps*, Penguin Books, London (a revised edition was published in 1986 by Allen & Unwin, London).

Graham, J. (1988) Post-modernism and Marxism, *Antipode*, Vol. 20, pp. 60–5.

Gregory, D. (1976) Rethinking historical geography, *Area*, Vol. 8, pp. 295–9.

Gregory, D. (1978) *Ideology, Science and Human Geography*, Hutchinson, London.

Gregory, D. (1980) The ideology of control: systems theory and geography, *Tijdschrift voor Economische en Sociale Geografie*, Vol. 71, pp. 327–42.

Gregory, D. (1981) Human agency and human geography, *Transactions of the Institute of British Geographers* (NS), Vol. 6, pp. 1–18.

Gregory, D. (1982a) A realist construction of the social, *Transactions of the Institute of British Geographers* (NS), Vol. 7, pp. 254–6.

Gregory, D. (1982b) *Regional Transformation and Industrial Revolution: A Geography of the Yorkshire Woollen Industry*, Macmillan, London.

Gregory, D. (1982c) Solid geometry: notes on the recovery of spatial structure, in P. Gould and G. Olsson (eds.) *A Search for Common Ground*, Pion, London.

Gregory, D. (1984) Space, time and politics in social theory: an interview with Anthony Giddens, *Environment and Planning D: Society and Space*, Vol. 2, pp. 123–32.

Gregory, D. (1985a) People, places and practices: the future of human geography, in R. King (ed.) *Geographical Futures*, Geographical Association, Sheffield.

Gregory, D. (1985b) Suspended animation: the stasis of diffusion theory, in D. Gregory and J. Urry (eds.), op. cit.

Gregory, D. (1986) Humanistic geography, Realism and Structuration theory, entries in R. J. Johnston, D. Gregory and D. M. Smith (eds.) *The Dictionary of Human Geography* (2nd edn), Blackwell, Oxford, pp. 207–10, pp. 387–90 and pp. 464–9.

Gregory, D. (1989a) Areal differentiation and post-modern human geography, in D. Gregory and R. Walford (eds.), op. cit.

Gregory, D. (1989b) Presences and absences: time-space relations and structuration theory, in D. Held and J. B. Thompson (eds.), op. cit.

Gregory, D. (1989c) The crisis of modernity? Human geography and critical social theory, in R. Peet and N. Thrift (eds.), op. cit., Vol. II.

Gregory, D. (1990a) *Chinatown*, Part Three? Soja and the missing spaces of social theory, *Strategies*.

Gregory, D. (1990b) Grand maps of history: structuration theory and social change, in J. Clark, C. Modgil and S. Modgil (eds.), op. cit.

Gregory, D. and Urry, J. (eds.) (1985) *Social Relations and Spatial Structures*, Macmillan, London.

Gregory, D. and Walford, R. (eds.) (1989) *Horizons in Human Geography*, Macmillan, London.

Gregson, N. (1986) On duality and dualism: the case of time-geography and structuration, *Progress in Human Geography*, Vol. 10, pp. 184–205.

Gregson, N. (1987a) Structuration theory: some thoughts on the possibilities for empirical research, *Environment and Planning D: Society and Space*, Vol. 5, pp. 73–91.

Gregson, N. (1987b) The CURS initiative: some further comments, *Antipode*, Vol. 19, pp. 364–370.

Gregson, N. (1989) On the (ir)relevance of structuration theory to empirical research, in D. Held and J. Thompson (eds.), op. cit.

Gross, D. (1982) Time-space relations in Giddens's social theory, *Theory, Culture and Society*, Vol. 1, pp. 83–7.

Guelke, L. (1974) An idealist alternative in human geography, *Annals of the Association of American Geographers*, Vol. 64, pp. 193–202.

Guelke, L. (1981) Idealism, in M. E. Harvey and B. P. Holly (eds.), op. cit.

Guelke, L. (1982) *Historical Understanding in Geography: An Idealist Approach*, Cambridge University Press.

Guelke, L. (1989a) Forms of life, history and mind: an idealist proposal for integrating perception and behaviour in human geography, in F. W. Boal and D. N. Livingstone (eds.) *The Behavioural Environment: Essays in Reflection, Application and Re-Evaluation*, Routledge, London.

Guelke, L. (1989b) Intellectual coherence and the foundations of geography, *Professional Geographer*, Vol. 41, pp. 123–130.

Habermas, J. (1972) *Knowledge and Human Interests*, Heinemann, London.

Habermas, J. (1976) *Legitimation Crisis*, Heinemann, London.

Habermas, J. (1981) Modernity versus postmodernity, *New German Critique*, no. 22, pp. 3–14.

Hägerstrand, T. (1970) What about people in regional science?, *Papers of the Regional Science Association*, Vol. 24, pp. 7–21.

Hägerstrand, T. (1974) Ecology under one perspective, in E. Bylund, H. Linderholm and O. Rune (eds.) *Ecological Problems of the Circumpolar North*, Norrbottens Museum, Lund.

Hägerstrand, T. (1975) Space, time and human conditions, in A. Karlqvist, L. Lundqvist and F. Snickars (eds.) *Dynamic Allocation of Urban Space*, Saxon House, Farnborough.

Hägerstrand, T. (1982) Diorama, path and project, *Tijdschrift voor Economische en Sociale Geografie*, Vol. 73, pp. 323–39.

Haggett, P. (1965) *Locational Analysis in Human Geography*, Edward Arnold, London.

Haggett, P. and Cliff, A. with Frey, A. (1977) *Locational Analysis in Human Geography* (2nd edn), Edward Arnold, London.

Haines-Young, R. and Petch, J. (1986) *Physical Geography: Its Nature and Methods*, Paul Chapman, London.

Hamnett, C. and McDowell, L. with Sarre, P. (eds.) (1989) *Restructuring Britain: The Changing Social Structure*, Open University/Sage, London.

Harré, R. (1979) *Social Being*, Blackwell, Oxford.

Harré, R. (1986) *Varieties of Realism*, Blackwell, Oxford.

Harrison, R. and Livingstone, D. N. (1979) There and back again: towards a critique of idealist human geography, *Area*, Vol. 11, pp. 75–9.

Hartshorne, R. (1939) *The Nature of Geography: A Critical Survey of Current Thought in the Light of the Past*, Association of American Geographers, Lancaster, Pa.

Hartshorne, R. (1955) 'Exceptionalism in geography' re-examined, *Annals of the Association of American Geographers*, Vol. 45, pp. 205–44.

Hartshorne, R. (1959) *Perspective on the Nature of Geography*, Rand McNally, Chicago, Ill.

Harvey, D. (1969a) Conceptual and measurement problems in the cognitive-behavioral approach to location theory, in K. R. Cox and R. G. Golledge (eds.) *Behavioral Problems in Geography: A Symposium*, Northwestern University Press, Evanston, Ill. (reprinted in K. R. Cox and R. G. Golledge (eds.) (1981) *Behavioural Problems in Geography Revisited*, Methuen, London).

Harvey, D. (1969b) *Explanation in Geography*, Edward Arnold, London.

Harvey, D. (1972) Revolutionary and counter-revolutionary theory in geography and the problem of ghetto formation, *Antipode*, Vol. 4, no. 2, pp. 1–13 (reprinted in *Antipode*, 1985, op. cit.).

Harvey, D. (1973) *Social Justice and the City*, Edward Arnold, London (a second edition was published in 1988 by Edward Arnold).

Harvey, D. (1974a) Population, resources and the ideology of science, *Economic Geography*, Vol. 50, pp. 256–77 (reprinted in S. Gale and G. Olsson (eds.) (1979) *Philosophy in Geography*, D. Reidel, Dordrecht).

Harvey, D. (1974b) What kind of geography for what kind of public policy?, *Transactions of the Institute of British Geographers*, Vol. 63, pp. 18–24.

Harvey, D. (1978) The urban process under capitalism: a framework for analysis, *International Journal of Urban and Regional Research*, Vol. 2, pp. 101–31.

Harvey, D. (1982) *The Limits to Capital*, Blackwell, Oxford.

Harvey, D. (1984) On the history and present condition of geography: an historical materialist manifesto, *Professional Geographer*, Vol. 36, pp. 1–11.

Harvey, D. (1987a) Flexible accumulation through urbanization: reflections on 'post-modernism' in the American city, *Antipode*, Vol. 19, pp. 260–86.

Harvey, D. (1987b) Three myths in search of a reality in urban studies, *Environment and Planning D: Society and Space*, Vol. 5, pp. 367–76.

Harvey, D. (1989a) From models to Marx: notes on the project to 'remodel' contemporary geography, in B. Macmillan (ed.), op. cit.

Harvey, D. (1989b) *The Condition of Postmodernity: An Enquiry into the Origins of Cultural Change*, Blackwell, Oxford.

Harvey, D. and Chatterjee, L. (1974) Absolute rent and the structuring of space by governmental and financial institutions, *Antipode*, Vol. 6, no. 1, pp. 22–36.

Harvey, D. and Scott, A. (1989) The practice of human geography: theory and empirical specificity in the transition from Fordism to flexible accumulation, in B. Macmillan (ed.), op. cit.

Harvey, M. E. and Holly, B. P. (eds.) (1981) *Themes in Geographic Thought*, Croom Helm, London.

Hasson, S. (1984) Humanistic geography from the perspective of Martin Buber's philosophy, *Professional Geographer*, Vol. 36, pp. 11–17.

Hay, A. (1985) Scientific method in geography, in R. J. Johnston (ed.), op. cit.

Hayford, A. M. (1974) The geography of women: an historical introduction, *Antipode*, Vol. 6, no. 2, pp. 1–19 (reprinted in *Antipode*, 1985, op. cit.).

Heidegger, M. (1962) *Being and Time*, Blackwell, Oxford.

Heidegger, M. (1971) *On the Way to Language*, Harper & Row, New York, NY.

Held, D. and Thompson, J. B. (eds.) (1989) *Social Theory of Modern Societies: Anthony Giddens and his Critics*, Cambridge University Press.

Herbertson, A. J. (1905) The major natural regions: an essay in systematic geography, *Geographical Journal*, Vol. 25, pp. 300–12.

Hill, M. H. (1985) Bound to the environment: towards a phenomenology of sightlessness, in D. Seamon and R. Mugerauer (eds.), op. cit.

Hill, M. R. (1981) Positivism: a 'hidden' philosophy in geography, in M. E. Harvey and B. P. Holly (eds.), op. cit.

Hirst, P. (1982) The social theory of Anthony Giddens: a new syncretism? *Theory, Culture and Society*, Vol. 1, pp. 78–82.

Horvarth, R. (1971) The 'Detroit Geographical Expedition and Institute' experience, *Antipode*, Vol. 3, no. 1, pp. 73–85.

Hudson, B. (1977) The new geography and the new imperialism, *Antipode*, Vol. 9, no. 2, pp. 12–19 (reprinted in *Antipode*, 1985, op. cit.).

Hudson, R. (1988) Uneven development in capitalist societies: changing spatial divisions of labour, forms of spatial organisation of production and service provision, and their impacts upon localities, *Transactions of the Institute of British Geographers* (NS), Vol. 13, pp. 484–96.

Hudson, R. (1989) Labour market changes and new forms of work in 'old' industrial regions: maybe flexibility for some but not flexible accumulation, *Environment and Planning D: Society and Space*, Vol. 7, pp. 5–30.

Huntington, E. (1915) *Civilization and Climate*, Yale University Press, New Haven, Conn.

Huntington, E. (1924) Geography and natural selection, *Annals of the Association of American Geographers*, Vol. 14, pp. 1–16.

Husserl, E. (1965) *Phenomenology and the Crisis of Philosophy* (reprints of 'Philosophy as rigorous science' (1911) and 'Philosophy and the crisis of European man' (1936)), Harper & Row, New York, NY.

Hutcheon, L. (1988) *A Poetics of Postmodernism: History, Theory and Fiction*, Routledge, London.

Huyssen, A. (1984) Mapping the postmodern, *New German Critique*, no. 33, pp. 5–52.

Institute for Workers Control (1977) A workers' enquiry into the motor industry, *Capital and Class*, Vol. 2, pp. 102–18.

Jackson, P. (1985) Urban ethnography, *Progress in Human Geography*, Vol. 9, pp. 157–76.

Jackson, P. (ed.) (1987) *Race and Racism: Essays in Social Geography*, Allen & Unwin, London.

Jackson, P. (1989) *Maps of Meaning: An Introduction to Cultural Geography*, Unwin Hyman, London.

Jackson, P. and Smith, S. J. (1984) *Exploring Social Geography*, Allen & Unwin, London.

James, P. E. (1952) Towards a fuller understanding of the regional concept, *Annals of the Association of American Geographers*, Vol. 42, pp. 195–222.

James, P. E. (1972) *All Possible Worlds: A History of Geographical Ideas*, Odyssey, Indianapolis, Ind.

Jameson, F. (1984) Post-modernism, or the cultural logic of late capitalism, *New Left Review*, no. 146, pp. 53–92.

Jansen, A. C. M. (1984) The atmosphere of a city centre, *Area*, Vol. 16, pp. 147–51.

Johnson, L. (1983) Bracketing lifeworlds: Husserlian phenomenology as a geographical method, *Australian Geographical Studies*, Vol. 21, pp. 102–8.

Johnson, L. (1987) (Un)realist perspectives: patriarchy and feminist challenges in geography, *Antipode*, Vol. 19, pp. 210–15.

Johnston, R. J. (ed.) (1985) *The Future of Geography*, Methuen, London.

Johnston, R. J. and Claval, P. (eds.) (1984) *Geography Since the Second World War: An International Survey*, Croom Helm, London.

Kearns, G. (1984) Closed space and political practice: Frederick Jackson Turner and Halford Mackinder, *Environment and Planning D: Society and Space*, Vol. 2, pp. 23–34.

Keat, R. and Urry, J. (1975) *Social Theory as Science*, Routledge & Kegan Paul, London; second edition published 1982.

Kellner, D. (1989) *Critical Theory, Marxism and Modernity*, Polity Press, Cambridge.

Kimble, G. H. T. (1951) The inadequacy of the regional concept, in L. D. Stamp and S. W. Wooldridge (eds.) *London Essays in Geography*, Longman, London.

Knox, P. L. (1987) The social production of the built environment: architects, architecture and the post-modern city, *Progress in Human Geography*, Vol. 11, pp. 354–77.

Kobayashi, A. and Mackenzie, S. (eds.) (1989) *Remaking Human Geography*, Unwin Hyman, London.

Kropotkin, P. (1978) Decentralisation, integration of labour and human education, reprinted in R. Peet (ed.), op. cit.

Kuhn, T. S. (1970) *The Structure of Scientific Revolutions* (2nd edn), University of Chicago Press, Chicago, Ill.

Lacoste, Y. (1973) An illustration of geographical warfare: bombing of the dikes on the Red River, North Vietnam, *Antipode*, Vol. 5, no. 2, pp. 1–13 (reprinted in R. Peet (ed.) (1978), op. cit.).

Lakatos, I. (1970) Falsification and the methodology of scientific research programmes, in I. Lakatos and A. Musgrave (eds.) *Criticism and the Growth of Knowledge*, Cambridge University Press.

Lash, S. and Urry, J. (1987) *The End of Organised Capitalism*, Polity Press, Cambridge.

Lauer, Q. (1965) Introduction, in E. Husserl, op. cit.

Le Corbusier, M. (1929) *The City of Tomorrow and its Planning*, Architectural Press, London.

Lefebvre, H. (1976) Reflections on the politics of space, *Antipode*, Vol. 8, no. 2, pp. 30–7 (reprinted in R. Peet (ed.) (1978), op. cit.).

Lewthwaite, G. R. (1966) Environmentalism and determinism: a search for clarification, *Annals of the Association of American Geographers*, Vol. 56, pp. 1–23.

Ley, D. (1974) The black inner city as frontier outpost: images and behavior of a Philadelphia neighborhood (Association of American Geographers, Monograph Series no. 7), Washington, DC.

Ley, D. (1977) Social geography and the taken-for-granted world, *Transactions of the Institute of British Geographers* (NS), Vol. 2, pp. 498–512 (reprinted in S. Gale and G. Olsson (eds.) (1979) *Philosophy in Geography*, D. Reidel, Dordrecht).

Ley, D. (1978) Social geography and social action, in D. Ley and M. S. Samuels (eds.), op. cit.

Ley, D. (1980a) Geography without man: a humanistic critique (University of Oxford, School of Geography Research Paper no. 24), Oxford.

Ley, D. (1980b) Liberal ideology and the postindustrial city, *Annals of the Association of American Geographers*, Vol. 70, pp. 238–58.

Ley, D. (1981a) Behavioural geography and the philosophies of meaning, in K. R. Cox and R. G. Golledge (eds.) *Behavioural Problems in Geography Revisited*, Methuen, London.

Ley, D. (1981b) Cultural/humanistic geography, *Progress in Human Geography*, Vol. 5, pp. 249–57.

Ley, D. (1982) Rediscovering man's place: commentary on Gregory's paper, *Transactions of the Institute of British Geographers* (NS), Vol. 7, pp. 248–53.

Ley, D. (1983) *A Social Geography of the City*, Harper & Row, New York, NY.

Ley, D. (1987) Styles of the times: liberal and neo-conservative landscapes in inner Vancouver, 1968–1986, *Journal of Historical Geography*, Vol. 13, pp. 40–56.

Ley, D. (1988) Interpretative social research in the inner city, in J. Eyles (ed.) *Research in Human Geography: Introductions and Investigations*, Blackwell, Oxford.

Ley, D. (1989) Fragmentation, coherence and limits to theory in human geography, in A. Kobayashi and S. Mackenzie (eds.), op. cit.

Ley, D. and Olds, K. (1988) Landscape as spectacle: world's fairs and the culture of heroic consumption, *Environment and Planning D: Society and Space*, Vol. 6, pp. 191–212.

Ley, D. and Samuels, M. S. (eds.) (1978a) *Humanistic Geography: Prospects and Problems*, Croom Helm, London.

Ley, D. and Samuels, M. S. (1978b) Introduction: contexts of modern humanism in geography, in D. Ley and M. S. Samuels (eds.), op. cit.

Lipietz, A. (1980) International regional polarisation and tertiarisation of society, *Papers of the Regional Science Association*, Vol. 44, pp. 3–17.

Lipietz, A. (1986) New tendencies in the international division of labour: regimes of accumulation and modes of regulation, in A. J. Scott and M. Storper (eds.), op. cit.

Little, J. and Peake, L. with Richardson, P. (eds.) (1988) *Women in Cities: Gender and the Urban Environment*, Macmillan, London.

Lovering, J. (1985) Regional intervention, defence industries and the structuring of space in Britain, *Environment and Planning D: Society and Space*, Vol. 3, pp. 85–107.

Lovering, J. (1989) Postmodernism, Marxism and locality research: the contribution of critical realism to the debate, *Antipode*, Vol. 21, pp. 1–12.

Lovering, J. (1990) Fordism's unknown successor: a comment on Scott's theory of flexible accumulation and the re-emergence of regional economies, *International Journal of Urban and Regional Research*, Vol. 14, pp. 159–74.

Lowenthal, D. (1961) Geography, experience and imagination: towards a geographical epistemology, *Annals of the Association of American Geographers*, Vol. 51, pp. 241–60.

Lowenthal, D. (ed.) (1967) *Environmental perception and behavior* (University of Chicago, Department of Geography, Research Paper no. 109), Chicago, Ill.

Lowenthal, D. and Bowden, M. J. (eds.) (1976) *Geographies of the Mind: Essays in Historical Geosophy*, Oxford University Press, New York, NY.

Lowther, G. R. (1959) Idealist history and historical geography, *Canadian Geographer*, Vol. 14, pp. 31–6.

Lyotard, J. F. (1984) *The Postmodern Condition: A Report on Knowledge*, Manchester University Press.

Mackenzie, S. (1989) *Visible Histories: Women and Environments in a Post-War British City*, McGill-Queen's University Press, London.

Mackinder, H. J. (1887) On the scope and methods of geography, *Proceedings of the Royal Geographical Society*, Vol. 9, pp. 141–60.

Macmillan, B. (ed.) (1989) *Remodelling Geography*, Blackwell, Oxford.

Mallory, W. E. and Simpson-Housley, P. (eds.) (1990) *Geography and Literature: A Meeting of the Disciplines*, University of Syracuse Press, Syracuse, NY.

Mandel, E. (1979) *Trotsky: A Study in the Dynamics of his Thought*, New Left Books, London.

Mann, M. (1986) *The Sources of Social Power, Volume 1: A History of Power from the Beginning to A.D. 1760*, Cambridge University Press.

Marcus, G. and Fischer, M. (eds.) (1986) *Anthropology as Cultural Critique: The Experimental Moment in the Human Sciences*, University of Chicago Press, Chicago, Ill.

Martin, A. (1951) The necessity for determinism, *Transactions of the Institute of British Geographers*, Vol. 17, pp. 1–11.

Marxism Today (1988) Special Issue on the New Times, October.

Massey, D. (1973) Towards a critique of industrial location theory, *Antipode*, Vol. 5, no. 3, pp. 33–9 (reprinted in R. Peet (ed.) (1978), op. cit.).

Massey, D. (1978) Regionalism: some current issues, *Capital and Class*, Vol. 6, pp. 106–25.

Massey, D. (1979) In what sense a regional problem?, *Regional Studies*, Vol. 13, pp. 233–43.

Massey, D. (1983) Industrial restructuring as class restructuring: production decentralisation and local uniqueness, *Regional Studies*, Vol. 17, pp. 73–90.

Massey, D. (1984) *Spatial Divisions of Labour: Social Structures and the Geography of Production*, Macmillan, London.

Massey, D. and Allen, J. (eds.) (1988) *Uneven Re-Development: Cities and Regions in Transition*, Open University/Hodder & Stoughton, London.

McDowell, L. (1983) Towards an understanding of the gender division of urban space, *Environment and Planning D: Society and Space*, Vol. 1, pp. 59–72.

McDowell, L. (1989a) Gender divisions, in C. Hamnett, L. McDowell and P. Sarre (eds.), op. cit.

McDowell, L. (1989b) Women, gender and the organisation of space, in D. Gregory and R. Walford (eds.), op. cit.

Mercer, D. and Powell, J. M. (1972) Phenomenology and related non-positivistic viewpoints in the social sciences (Monash University Publications in Geography no. 1), Melbourne.

Mills, C. A. (1988) 'Life on the upslope': the postmodern landscape of gentrification, *Environment and Planning D: Society and Space*, Vol. 6, pp. 169–90.

Mills, W. J. (1982) Positivism reversed: the relevance of Giambattista Vico, *Transactions of the Institute of British Geographers* (NS), Vol. 7, pp. 1–14.

Moos, A. I. and Dear, M. J. (1986) Structuration theory in urban analysis: 1. theoretical exegesis, *Environment and Planning A*, Vol. 18, pp. 231–52.

Morrill, R. (1969/1970) Geography and the transformation of society, *Antipode*, Vol. 1, no. 1, pp. 6–9 and Vol. 2, no. 1, pp. 4–10 (reprinted in *Antipode*, 1985, op. cit. Page nos. in text refer to this reprint).

Mugerauer, R. (1981) Concerning regional geography as a hermeneutical discipline, *Geographische Zeitschrift*, Vol. 69, pp. 57–67.

Mugerauer, R. (1984) Mapping the movement of geographical inquiry: a commentary, in T. F. Saarinen, D. Seamon and J. L. Sell (eds.), op. cit.

Mugerauer, R. (1985) Language and the emergence of the environment, in D. Seamon and R. Mugerauer (eds.), op. cit.

Mumford, L. (1960) Universal city, in C. H. Kraeling and R. M. Adams (eds.) *City Invisible*, University of Chicago Press, Chicago, Ill.

Murgatroyd, L. (1989) Only half the story: some blinkering effects of 'malestream' sociology, in D. Held and J. B. Thompson (eds.), op. cit.

Ogborn, M. (1990) 'A lynx-eyed and iron-handed system': the state regulation of prostitution in nineteenth-century Britain (unpublished paper).

O'Hagan, S. (1990) Pop's high priest of sex, *Sunday Correspondent*, 17 June.

Olsson, G. (1972) Some notes on geography and social engineering, *Antipode*, Vol. 4, no. 1, pp. 1–22.

Olsson, G. (1974a) Servitude and inequality in spatial planning: ideology and methodology in conflict, *Antipode*, Vol. 6, no. 1, pp. 16–21 (reprinted in R. Peet (ed.) (1978), op. cit.).

Olsson, G. (1974b) The dialectics of spatial analysis, *Antipode*, Vol. 6, no. 3, pp. 50–62.

Olsson, G. (1980) *Birds in Egg: Eggs in Bird*, Pion, London.

Olsson, G. (1987) The social space of silence, *Environment and Planning D: Society and Space*, Vol. 5, pp. 249–62.

Openshaw, S. (1984) The modifiable areal unit problem (CATMOG no. 38), Norwich.

Outhwaite, W. (1983) *Concept Formation in Social Science*, Routledge & Kegan Paul, London.

Outhwaite, W. (1987) *New Philosophies of Science: Realism, Hermeneutics and Critical Theory*, Macmillan, London.

Outhwaite, W. (1989) 'Lost and found': review of Bhaskar's *Reclaiming Reality, The Times Higher Education Supplement*, 15 September.

Owens, C. (1985) The discourse of others: feminists and postmodernism, in H. Foster (ed.) *Postmodern Culture*, Pluto, London.

Parkes, D. and Thrift, N. (1980) *Time, Spaces and Places*, Wiley, Chichester.

Paterson, J. H. (1974) Writing regional geography: problems and progress in the Anglo-American realm, *Progress in Geography*, Vol. 6, pp. 1–26.

Peet, R. (1975) Inequality and poverty: a Marxist-geographic theory, *Annals of the Association of American Geographers*, Vol. 65, pp. 564–71 (reprinted in R. Peet (ed.) (1978), op. cit.).

Peet, R. (1977) The development of radical geography in the United States, *Progress in Human Geography*, Vol. 1, pp. 240–63 (reprinted in R. Peet (ed.) (1978), op. cit.).

Peet, R. (ed.) (1978) *Radical Geography: Alternative Viewpoints on Contemporary Social Issues*, Methuen, London.

Peet, R. (1981) Spatial dialectics and Marxist geography, *Progress in Human Geography*, Vol. 5, pp. 105–10.

Peet, R. (1985a) Radical geography in the United States: a personal history, *Antipode*, Vol. 17, nos. 2 and 3, pp. 1–7.

Peet, R. (1985b) The social origins of environmental determinism, *Annals of the Association of American Geographers*, Vol. 75, pp. 309–33.

Peet, R. and Thrift, N. (eds.) (1989) *New Models in Geography, Volumes I and II*, Unwin Hyman, London.

Perry, P. J. (1990) S. W. Wooldridge: the geographer as humanist, *Transactions of the Institute of British Geographers* (NS), Vol. 15, pp. 227–31.

Philo, C. (1984) Reflections on Gunnar Olsson's contribution to the discourse of contemporary human geography, *Environment and Planning D: Society and Space*, Vol. 2, pp. 217–40.

Philo, C. (1986) 'The same and the other': on geographies, madness and outsiders (Loughborough University of Technology, Department of Geography, Occasional Paper no. 11), Loughborough.

Philo, C. (1989a) 'Enough to drive one mad': the organisation of space in nineteenth-century lunatic asylums, in J. Wolch and M. Dear (eds.) *The Power of Geography: How Territory Shapes Social Life*, Unwin Hyman, London.

Philo, C. (1989b) Thoughts, words and 'creative locational acts', in F. W. Boal and D. N. Livingstone (eds.) *The Behavioural Environment: Essays in Reflection, Application and Re-Evaluation*, Routledge, London.

Philo, C. (1991a) Foucault's geography, forthcoming, in *Environment and Planning D: Society and Space*.

Philo, C. (1991b) History, geography and the 'still greater mystery of historical geography', forthcoming, in D. Gregory, R. Martin and G. Smith (eds.) *Rethinking Human Geography: Society, Space and the Social Sciences*, Macmillan, London.

Pickles, J. (1985) *Phenomenology, Science and Geography: Spatiality and the Human Sciences*, Cambridge University Press.

Pocock, D. C. D. (ed.) (1981a) *Humanistic Geography and Literature: Essays on the Experience of Place*, Croom Helm, London.

Pocock, D. C. D. (1981b) Introduction: imaginative literature and the geographer, in D. C. D. Pocock (ed.), op. cit.

Pocock, D. C. D. (ed.) (1988a) Humanistic approaches in geography (University of Durham, Department of Geography, Occasional Paper no. 22), Durham.

Pocock, D. C. D. (1988b) The music of geography, in D. C. D. Pocock (ed.), op. cit.

Pollert, A. (1988) Dismantling flexibility, *Capital and Class*, Vol. 34, pp. 42–75.

Poole, R. (1988) Deconstruction, entry in A. Bullock, O. Stallybrass and S. Trombley (eds.) *The Fontana Dictionary of Modern Thought*, Fontana, London, pp. 205–6.

Popper, K. R. (1959) *The Logic of Scientific Discovery*, Hutchinson, London.

Pred, A. (1967) Behaviour and location: foundations for a geographic and dynamic location theory, part I (Lund Studies in Geography, Series B: Human Geography, Paper no. 27), Gleerup, Lund.

Pred, A. (1977) The choreography of existence: comments on Hägerstrand's time-geography and its usefulness, *Economic Geography*, Vol. 53, pp. 207–21.

Pred, A. (1978) The impact of technological and institutional innovations of life content: some time-geographic observations, *Geographical Analysis*, Vol. 10, pp. 345–72.

Pred, A. (1981) Social reproduction and the time-geography of everyday life, *Geografiska Annaler*, Vol. 63B, pp. 5–22.

Pred, A. (1981) Social reproduction and the time-geography of everyday life, *Geografiska Annaler*, Vol. 63B, pp. 5–22.

Pred, A. (1983) Structuration and place: on the becoming of sense of place and structure of feeling, *Journal for the Theory of Social Behaviour*, Vol. 13, pp. 45–68.

Pred, A. (1984a) Place as historically-contingent process: structuration and the time-geography of becoming places, *Annals of the Association of American Geographers*, Vol. 74, pp. 279–97.

Pred, A. (1984b) Structuration, biography formation and knowledge: observations on port growth during the mercantile period, *Environment and Planning D: Society and Space*, Vol. 2, pp. 251–75.

Pred, A. (1985) The social becomes the spatial, the spatial becomes the social: enclosures, social change and the becoming of place in the Swedish province of Skåne, in D. Gregory and J. Urry (eds.), op. cit.

Pred, A. (1987) *Place, Practice and Structure: The Emergence and Aftermath of Enclosures in the Plains Villages of Southwestern Skåne, 1750–1850*, Cambridge University Press.

Pred, A. (1989a) *Lost Words and Lost Worlds: Modernity and the Language of Everyday Life in Late Nineteenth-Century Stockholm*, Cambridge University Press.

Pred, A. (1989b) The locally spoken word and local struggles, *Environment and Planning D: Society and Space*, Vol. 7, pp. 211–33.

Pred, A. (1990) *Making Histories and Constructing Human Geographies*, Westview Press, Boulder, Colorado.

Prince, H. (1971) Questions of social relevance, *Area*, Vol. 3, pp. 150–3.

Quaini, M. (1982) *Geography and Marxism*, Blackwell, Oxford (includes a bibliography of radical geographical literature written in English).

Quinton, A. (1988) Epistemology, Existentialism and Ontology, entries in A. Bullock, O. Stallybrass and S. Trombley (eds.) *The Fontana Dictionary of Modern Thought* (2nd edn), Fontana, London, p. 279, pp. 296–7 and pp. 605–6.

Quoniam, S. (1988) A painter, geographer of Arizona, *Environment and Planning D: Society and Space*, Vol. 6, pp. 3–14.

Raban, J. (1974) *Soft City*, Hamish Hamilton, London.

Relph, E. (1970) An inquiry into the relations between phenomenology and geography, *Canadian Geographer*, Vol. 14, pp. 193–201.

Relph, E. (1976) *Place and Placelessness*, Pion, London.

Relph, E. (1977) Humanism, phenomenology and geography: commentary on Entrikin's paper, *Annals of the Association of American Geographers*, Vol. 67, pp. 177–9.

Relph, E. (1981a) Phenomenology, in M. E. Harvey and B. P. Holly (eds.), op. cit.

Relph, E. (1981b) *Rational Landscapes and Humanistic Geography*, Croom Helm, London.

Relph, E. (1985) Geographical experiences and being-in-the-world: the phenomenological origins of geography, in D. Seamon and R. Mugerauer (eds.), op. cit.

Relph, E. (1989) A curiously unbalanced condition of the powers of the mind: realism and the ecology of environmental experience, in F. W. Boal and D. N. Livingstone (eds.) *The Behavioural Environment: Essays in Reflection, Application and Re-Evaluation*, Routledge, London.

Richards, P. (1974) Kant's geography and mental maps, *Transactions of the Institute of British Geographers*, Vol. 61, pp. 1–16.

Richardson, M. (1981) On 'the Superorganic in American cultural geography': commentary on Duncan's paper, *Annals of the Association of American Geographers*, Vol. 71, pp. 284–7.

Robson, B. (1986) Research issues in the changing urban and regional system, *Regional Studies*, Vol. 20, pp. 203–7.

Rodaway, P. (1988) Opening environmental experience, in D. C. D. Pocock (ed.), op. cit.

Rorty, R. (1980) *Philosophy and the Mirror of Nature*, Blackwell, Oxford.

Rorty, R. (1982) *Consequences of Pragmatism*, University of Minnesota Press, Minneapolis, Minn.

Rose, C. (1981) Wilhelm Dilthey's philosophy of human understanding: a neglected heritage of contemporary humanistic geography, in D. R. Stoddart (ed.) *Geography, Ideology and Social Concern*, Blackwell, Oxford.

Rose, G. (1988) Locality, politics and culture: Poplar in the 1920s, *Environment and Planning D: Society and Space*, Vol. 6, pp. 151–68.

Rowles, G. (1978a) Reflections on experimental field work, in D. Ley and M. Samuels (eds.), op. cit.

Rowles, G. (1978b) *The Prisoners of Space? Exploring the Geographical Experience of Older People*, Westview Press, Boulder, Colo.

Saarinen, T. F. and Seamon D. with Sell J. L. (eds.) (1984) *Environmental perception and behavior: an inventory and prospect* (University of Chicago, Department of Geography, Research Paper no. 209), Chicago, Ill.

Sack, R. D. (1980) *Conceptions of Space in Social Thought: A Geographic Perspective*, Macmillan, London.

Samuels, M. (1978) Existentialism and human geography, in D. Ley and M. S. Samuels (eds.), op. cit.

Samuels, M. (1979) The biography of landscape: cause and culpability, in D. W. Meinig (ed.) *The Interpretation of Ordinary Landscapes: Geographical Essays*, Oxford University Press, New York, NY.

Samuels, M. (1981) An existential geography, in M. E. Harvey and B. P. Holly (eds.), op. cit.

Santos, M. (1974) Geography, Marxism and underdevelopment, *Antipode*, Vol. 6, no. 3, pp. 1–9.

Santos, M. (1977) Spatial dialectics: the two circuits of urban economy in underdeveloped countries, *Antipode*, Vol. 9, no. 3, pp. 49–60 (reprinted in *Antipode*, 1985, op. cit.).

Sarre, P. (1987) Realism in practice, *Area*, Vol. 19, pp. 3–10.

Sartre, J.-P. (1948) *Existentialism and Humanism*, Methuen, London.

Sauer, C. O. (1924) The survey method in geography and its objectives, *Annals of the Association of American Geographers*, Vol. 14, pp. 17–33.

Sauer, C. O. (1956) The agency of man on earth, in W. L. Thomas (ed.) *Man's Role in Changing the Face of the Earth, Volume 1*, University of Chicago Press, Chicago, Ill.

Saunders, P. (1989) Space, urbanism and the created environment, in D. Held and J. B. Thompson (eds.), op. cit.

Saunders, P. and Williams, P. (1986) The new conservatism: some thoughts on recent and future developments in urban studies, *Environment and Planning D: Society and Space*, Vol. 4, pp. 393–9.

Sayer, A. (1978) Mathematical modelling in regional science and political economy: some comments, *Antipode*, Vol. 10, no. 2, pp. 79–86 (reprinted in *Antipode*, 1985, op. cit.).

Sayer, A. (1982) Explanation in economic geography, *Progress in Human Geography*, Vol. 6, pp. 68–88.

Sayer, A. (1983) Review of Giddens's *A Contemporary Critique of Historical Materialism, Volume 1, Environment and Planning D: Society and Space*, Vol. 1, pp. 109–14.

Sayer, A. (1984) *Method in Social Science: A Realist Approach*, Hutchinson, London.

Sayer, A. (1985a) Realism and geography, in R. J. Johnston (ed.), op. cit.

Sayer, A. (1985b) The difference that space makes, in D. Gregory and J. Urry (eds.), op. cit.

Sayer, A, (1987) Hard work and its alternatives, *Environment and Planning D: Society and Space*, Vol. 5, pp. 395–9.

Sayer, A. (1989a) On the dialogue between humanism and historical materialism in geography, in A. Kobayashi and S. Mackenzie (eds.), op. cit.

Sayer, A. (1989b) Post–Fordism in question, *International Journal of Urban and Regional Research*, Vol. 13, pp. 666–95.

Sayer, A. (1989c) The 'new' regional geography and problems of narrative, *Environment and Planning D: Society and Space*, Vol. 7, pp. 253–76.

Schaefer, F. (1953) Exceptionalism in geography: a methodological examination, *Annals of the Association of American Geographers*, Vol. 43, pp. 226–49.

Schiller, F. C. S. (1907) *Studies in Humanism*, Macmillan, London.

Schmitt, C. B. (1988) *The Cambridge History of Renaissance Philosophy*, Cambridge University Press.

Scott, A. J. (1988a) Flexible production systems and regional development: the rise of new industrial spaces in North America and western Europe, *International Journal of Urban and Regional Research*, Vol. 12, pp. 171–86.

Scott, A. J. (1988b) *New Industrial Spaces*, Pion, London.

Scott, A. J. and Storper M. (eds.) (1986) *Production, Work, Territory: The Geographical Anatomy of Industrial Capitalism*, Allen & Unwin, London.

Seamon, D. (1978) Goethe's approach to the natural world: implications for environmental theory and education, in D. Ley and M. S. Samuels (eds.), op. cit.

Seamon, D. (1979) *A Geography of the Lifeworld: Movement, Rest and Encounter*, Croom Helm, London.

Seamon, D. (1980) Body, subject, time-space routines and place ballets, in A. Buttimer and D. Seamon (eds.), op. cit.

Seamon, D. (1984) Philosophical directions in behavioral geography with an emphasis on the phenomenological contribution, in T. F. Saarinen, D. Seamon and J. L. Sell (eds.), op. cit.

Seamon, D. (1987) Phenomenology and environment-behavior research, in E. H. Zube and G. T. Moore (eds.) *Advances in Environment, Behavior and Design*, Plenum Press, New York, NY.

Seamon, D. (1989) Humanistic and phenomenological advances in environmental design, *Humanistic Psychologist*, Vol. 17, pp. 280–93.

Seamon, D. and Mugerauer, R. (eds.) (1985a) *Dwelling, Place and Environment: Towards a Phenomenology of Person and World*, Martinus Nijhoff, Dordrecht (reprinted in 1989 by Columbia University Press. New York, NY).

Seamon, D. and Mugerauer, R. (1985b) Dwelling, place and environment, in D. Seamon and R. Mugerauer (eds.), op. cit.

Seamon, D. and Nordin, C. (1980) Marketplace as place ballet: a Swedish example, *Landscape*, Vol. 24, pp. 35–41.

Semple, E. C. (1911) *Influences of Geographic Environment: On the Basis of Ratzel's System of Anthropo-Geography*, Constable, London.

Semple, E. C. (1915) The barrier boundary of the Mediterranean basin and its northern breaches as factors in history, *Annals of the Association of American Geographers*, Vol. 5, pp. 27–59.

Shields, R. (1988) Social spatialisation and the built environment: the West Edmonton Mall, *Environment and Planning D: Society and Space*, Vol. 7, pp. 147–64.

Short, J. R. (1989) Yuppies, yuffies and the new urban order, *Transactions of the Institute of British Geographers* (NS), Vol. 14, pp. 173–88.

Sibley, D. (1981a) *Outsiders in Urban Societies*, Blackwell, Oxford.

Sibley, D. (1981b) The notion of order in spatial analysis, *Professional Geographer*, Vol. 33, pp. 1–5.

Sibley, D. (1981c) Reply to Richard Walker, *Professional Geographer*, Vol. 33, pp. 10–11.

Slater, D. (1977) Geography and underdevelopment, *Antipode*, Vol. 9, no. 3, pp. 1–31.

Smith, D. (1982) 'Put not your trust in Princes': a commentary upon Anthony Giddens and the absolute state, *Theory, Culture and Society*, Vol. 1, pp. 93–7.

Smith, D. M. (1971) Radical geography: the next revolution?, *Area*, Vol. 3, pp. 153–7.

Smith, D. M. (1977) *Human Geography: A Welfare Approach*, Edward Arnold, London.

Smith, D. M. (1988) Towards an interpretative human geography, in J. Eyles and D. M. Smith (eds.), op. cit.

Smith, N. (1979) Geography, science and post-positivist modes of explanation, *Progress in Human Geography*, Vol. 3, pp. 356–83.

Smith, N. (1981) Degeneracy in theory and practice: spatial interactionism and radical eclecticism, *Progress in Human Geography*, Vol. 5, pp. 111–18.

Smith, N. (1984) *Uneven Development: Nature, Capital and the Production of Space*, Blackwell, Oxford.

Smith, N. (1987a) Academic war over the field of geography: the elimination of geography at Harvard, 1947–1951, *Annals of the Association of American Geographers*, Vol. 77, pp. 155–72.

Smith, N. (1987b) Dangers of the empirical turn: some comments on the CURS initiative, *Antipode*, Vol. 19, pp. 59–68.

Smith, N. (1987c) Rascal concepts, minimalising discourse and the politics of geography, *Environment and Planning D: Society and Space*, Vol. 5, pp. 377–83.

Smith, N. (1988) The region is dead! Long live the region!, *Political Geography Quarterly*, Vol. 7, pp. 141–52.

Smith, N. (1989) Uneven development and location theory: towards a synthesis, in R. Peet and N. Thrift (eds.), op. cit., Vol. I.

Smith, N. and Williams, P. (eds.) (1986) *Gentrification of the City*, Unwin Hyman, London.

Smith, S. J. (1981) Humanistic method in contemporary social geography, *Area*, Vol. 13, pp. 293–8.

Smith, S. J. (1984) Practicing humanistic geography, *Annals of the Association of American Geographers*, Vol. 74, pp. 353–74.

Smith, S. J. (1988) Constructing local knowledge: the analysis of self in everyday life, in J. Eyles and D.M. Smith (eds.), op. cit.

Society and Space (1988) Special issue on the new geography and sociology of production, *Environment and Planning D: Society and Space*, Vol. 6, no. 3.

Soja, E. W. (1980) The socio-spatial dialectic, *Annals of the Association of American Geographers*, Vol. 70, pp. 207–25.

Soja, E. W. (1985) The spatiality of social life: towards a transformative re-theorisation, in D. Gregory and J. Urry (eds.), op. cit.

Soja, E. W. (1986) Taking Los Angeles apart: some fragments of a critical human geography, *Environment and Planning D: Society and Space*, Vol. 4, pp. 255–72 (reprinted with minor changes in E. W. Soja (1989), op. cit.).

Soja, E. W. (1989) *Postmodern Geographies: The Reassertion of Space in Critical Social Theory*, Verso, London.

Soja, E. W. and Hadjimachalis, C. (1979) Between geographical materialism and spatial fetishism: some observations on the development of Marxist spatial analysis, *Antipode*, Vol. 11, no. 3, pp. 3–11 (reprinted in *Antipode*, 1985, op. cit.).

Spate, O. H. K. (1960) Quantity and quality in geography, *Annals of the Association of American Geographers*, Vol. 50, pp. 377–94.

Stallybrass, P. and White, A. (1986) *The Politics and Poetics of Transgression*, Methuen, London.

Stamp, L. D. (1957) Geographical agenda: a review of some tasks awaiting geographical attention, *Transactions of the Institute of British Geographers*, Vol. 23, pp. 1–17.

Stephenson, D. (1974) The Toronto Geographical Expedition, *Antipode*, Vol. 6, no. 2, pp. 98–101.

Stoddart, D. R. (1966) Darwin's impact on geography, *Annals of the Association of American Geographers*, Vol. 56, pp. 683–98 (reprinted with minor changes in D. R. Stoddart (1986), op. cit.).

Stoddart, D. R. (1967) Growth and structure of geography, *Transactions of the Institute of British Geographers*, Vol. 41, pp. 1–19.

Stoddart, D. R. (1986) *On Geography and its History*, Blackwell, Oxford.

Stokes, E. (1987) Maori geography or geography of Maoris, *New Zealand Geographer*, Vol. 43, pp. 118–23.

Storper, M. (1985) The spatial and temporal constitution of social action: a critical reading of Giddens, *Environment and Planning D: Society and Space*, Vol. 3, pp. 407–24.

Storper, M. and Christopherson, S. (1987) Flexible specialization and regional industrial agglomeration: the case of the US motion picture industry, *Annals of the Association of American Geographers*, Vol. 77, pp. 104–17.

Suttles, G. D. (1968) *The Social Order of the Slum*, Chicago University Press, Chicago, Ill.

Taylor, P. J. (1976) An interpretation of the quantification debate in British geography, *Transactions of the Institute of British Geographers* (NS), Vol. 1, pp. 129–42.

Taylor, P. J. (1989) *Political Geography: World-Economy, Nation-State and Locality* (2nd edn), Longman, Harlow.

Thompson, E. P. (1963) *The Making of the English Working Class*, Penguin Books, Harmondsworth.

Thompson, E. P. (1965) Peculiarities of the English, in R. Milliband and J. Saville (eds.) *The Socialist Register*, 1965, Merlin Press, London.

Thompson, J. B. (1984) *Studies in the Theory of Ideology*, Polity Press, Cambridge.

Thrift, N. (1977) *An Introduction to Time-Geography: Concepts and Techniques in Modern Geography No. 13*, Geo-Abstracts, Norwich.

Thrift, N. (1983) On the determination of social action in space and time, *Environment and Planning D: Society and Space*, Vol. 1, pp. 23–57.

Thrift, N. (1985) Bear and mouse or bear and tree? Anthony Giddens's reconstitution of social theory, *Sociology*, Vol. 19, pp. 609–23.

Thrift, N. and Pred, A. (1981) Time-geography: a new beginning, *Progress in Human Geography*, Vol. 5, pp. 277–86.

Tivers, J. (1985) *Women Attached: The Daily Lives of Women with Young Children*, Croom Helm, London.

Tivers, J. (1988) Women with young children: constraints on activities in the urban environment, in J. Little, L. Peake and P. Richardson (eds.), op. cit.

Touraine, A. (1977) *The Self-Production of Society*, Chicago University Press, Chicago, Ill.

Tuan, Y.-F. (1971) Geography, phenomenology and the study of human nature, *Canadian Geographer*, Vol. 15, pp. 181–92.

Tuan, Y.-F. (1974a) Space and place: humanistic perspective, *Progress in Geography*, Vol. 6, pp. 211–52.

Tuan, Y.-F. (1974b) *Topophilia: A Study of Environmental Perception, Attitudes and Values*, Prentice-Hall, Englewood Cliffs, NJ.

Tuan, Y.-F. (1976) Humanistic geography, *Annals of the Association of American Geographers*, Vol. 66, pp. 266–76.

Tuan, Y.-F. (1977) *Space and Place: The Perspective of Experience*, Edward Arnold, London.

Tuan, Y.-F. (1979) *Landscapes of Fear*, Blackwell, Oxford.

Urry, J. (1981) Localities, regions and social class, *International Journal of Urban and Regional Research*, Vol. 5, pp. 455–74.

Urry, J. (1982) Duality of structure: some critical issues, *Theory, Culture and Society*, Vol. 1, pp. 100–6.

Urry, J. (1983) Realism and the analysis of space, *International Journal of Urban and Regional Research*, Vol. 7, pp. 122–7.

Urry, J. (1985) Social relations, space and time, in D. Gregory and J. Urry (eds.), op. cit.

Vidal de la Blache, P. (1917) *La France de L'Est*, Paris.

Vidal de la Blache, P. (1926) *Principles of Human Geography*, Constable, London.

Walker, R. (1981) Left-wing libertarianism, an academic disorder: a response to David Sibley, *Professional Geographer*, Vol. 33, pp. 5–9.

Walker, R. (1989) What's left to do?, *Antipode*, Vol. 21, pp. 133–65.

Walker, R. and Greenberg, D. A. (1982) Post-industrialism and political reform in the city: a critique, *Antipode*, Vol. 14, no. 1, pp. 17–32.

Wallace, I. (1978) Towards a humanised conception of economic geography, in D. Ley and M. S. Samuels (eds.), op. cit.

Walmsley, D. J. (1974) Positivism and phenomenology in human geography, *Canadian Geographer*, Vol. 18, pp. 95–107.

Warde, A. (1989) Recipes for a pudding: a comment on locality, *Antipode*, Vol. 21, pp. 274–81.

Watts, S. J. and Watts, S. J. (1978) On the idealist alternative in geography and history, *Professional Geographer*, Vol. 30, pp. 123–7.

Western, J. (1981) *Outcast Cape Town*, Allen & Unwin, London.

White, G. and Murray, R. with White, C. (eds.) (1983) *Revolutionary Socialist Development in the Third World*, Harvester, Brighton.

Whitehand, J. W. R. (1990) An assessment of 'progress', *Progress in Human Geography*, Vol. 14, pp. 12–23.

Whittlesey, D. (1954) The regional concept and the regional method, in P. E. James and C. F. Jones (eds.) *American Geography: Inventory and Prospect*, Syracuse University Press, Syracuse, NY.

Wiles, P. (ed.) (1982) *The New Communist Third World*, Croom Helm, London.

Williams, S. (1981) Realism, Marxism and human geography, *Antipode*, Vol. 13, no. 2, pp. 31–8.

Wilson, A. G. (1972) Theoretical geography: some speculations, *Transactions of the Institute of British Geographers*, Vol. 57, pp. 31–44.

Winchester, H. P. and White, P. E. (1988) The location of marginalised groups in the inner city, *Environment and Planning D: Society and Space*, Vol. 6, pp. 37–54.

Wolpert, J. (1964) The decision process in spatial context, *Annals of the Association of American Geographers*, Vol. 54, pp. 337–58.

Women and Geography Study Group of the IBG (1984) *Geography and Gender: An Introduction to Feminist Geography*, Hutchinson, London.

Wooldridge, S. W. (1969a) On taking the 'ge-' out of 'geography', reprinted in S. W. Wooldridge (1969), op. cit.

Wooldridge, S. W. (1969b) *The Geographer as Scientist: Essays on the Scope and Nature of Geography*, Greenwood Press, New York, NY.

Wright, E. (1983) Giddens's critique of Marxism, *New Left Review*, no. 138, pp. 11–35.

Wright, J. K. (1925) The geographical lore of the time of the Crusades: a study in the history of Medieval science and tradition in Western Europe (American Geographical Society of New York, Research Paper no. 15), New York, NY (reprinted in 1965 by Dover Publications).

Wright, J. K. (1947) *Terrae incognitae*: the place of imagination in geography, *Annals of the Association of American Geographers*, Vol. 37, pp. 1–15 (reprinted in J. K. Wright (1966) *Human Nature in Geography: Fourteen Papers, 1925–1965*, Harvard University Press, Cambridge, Mass.).

AUTHOR INDEX

SUBJECT INDEX